# Millimeter Wave and Terahertz Source, Sensing and Imaging

# Millimeter Wave and Terahertz Source, Sensing and Imaging

Guest Editors

**Yubing Gong**
**Min Hu**
**Xuyuan Chen**
**Xuequan Chen**

Basel • Beijing • Wuhan • Barcelona • Belgrade • Novi Sad • Cluj • Manchester

*Guest Editors*

Yubing Gong
School of Electronic Science
and Engineering
University of Electronic
Science and Technology
of China
Chengdu
China

Min Hu
School of Electronic Science
and Engineering
University of Electronic
Science and Technology
of China
Chengdu
China

Xuyuan Chen
Department of Microsystems
University of South-Eastern
Norway
Tønsberg
Norway

Xuequan Chen
GBA Branch of Aerospace
Information Research
Institute
Chinese Academy of Sciences
Guangzhou
China

*Editorial Office*
MDPI AG
Grosspeteranlage 5
4052 Basel, Switzerland

This is a reprint of the Special Issue, published open access by the journal *Sensors* (ISSN 1424-8220), freely accessible at: https://www.mdpi.com/journal/sensors/special_issues/N2B32NTAR8.

For citation purposes, cite each article independently as indicated on the article page online and as indicated below:

Lastname, A.A.; Lastname, B.B. Article Title. *Journal Name* **Year**, *Volume Number*, Page Range.

ISBN 978-3-7258-3521-8 (Hbk)
ISBN 978-3-7258-3522-5 (PDF)
https://doi.org/10.3390/books978-3-7258-3522-5

© 2025 by the authors. Articles in this book are Open Access and distributed under the Creative Commons Attribution (CC BY) license. The book as a whole is distributed by MDPI under the terms and conditions of the Creative Commons Attribution-NonCommercial-NoDerivs (CC BY-NC-ND) license (https://creativecommons.org/licenses/by-nc-nd/4.0/).

# Contents

**Kuangyi Xu, Zachery B. Harris, Paul Vahey and M. Hassan Arbab**
THz Polarimetric Imaging of Carbon Fiber-Reinforced Composites Using the Portable Handled Spectral Reflection (PHASR) Scanner
Reprinted from: *Sensors* **2024**, *24*, 7467, https://doi.org/10.3390/s24237467 . . . . . . . . . . . . . 1

**Chuanhong Xiao, Ruizhe Ren, Zhenhua Wu, Yijun Li, Qing You, Zongjun Shi, et al.**
Research of a 0.14 THz Dual-Cavity Parallel Structure Extended Interaction Oscillator
Reprinted from: *Sensors* **2024**, *24*, 5891, https://doi.org/10.3390/s24185891 . . . . . . . . . . . . . 12

**Xingxing Xu, Fu Tang, Xiaoqiuyan Zhang and Shenggang Liu**
Unveiling the Terahertz Nano-Fingerprint Spectrum of Single Artificial Metallic Resonator
Reprinted from: *Sensors* **2024**, *24*, 5866, https://doi.org/10.3390/s24185866 . . . . . . . . . . . . . 21

**Duo Xu, Tenglong He, Yuan Zheng, Zhigang Lu, Huarong Gong, Zhanliang Wang, et al.**
A Symmetrical Quasi-Synchronous Step-Transition Folded Waveguide Slow Wave Structure for 650 GHz Traveling Wave Tubes
Reprinted from: *Sensors* **2024**, *24*, 5289, https://doi.org/10.3390/s24165289 . . . . . . . . . . . . . 34

**Yuxin Wang, Jingyu Guo, Yang Dong, Duo Xu, Yuan Zheng, Zhigang Lu, et al.**
A Staggered Vane-Shaped Slot-Line Slow-Wave Structure for W-Band Dual-Sheet Electron-Beam-Traveling Wave Tubes
Reprinted from: *Sensors* **2024**, *24*, 3709, https://doi.org/10.3390/s24123709 . . . . . . . . . . . . . 46

**Yang Cao, Kathirvel Nallappan, Guofu Xu and Maksim Skorobogatiy**
Resonant Gas Sensing in the Terahertz Spectral Range Using Two-Wire Phase-Shifted Waveguide Bragg Gratings
Reprinted from: *Sensors* **2023**, *23*, 8527, https://doi.org/10.3390/s23208527 . . . . . . . . . . . . . 55

**Zechuan Wang, Junwan Zhu, Zhigang Lu, Jingrui Duan, Haifeng Chen, Shaomeng Wang, et al.**
A Novel Staggered Double-Segmented Grating Slow-Wave Structure for 340 GHz Traveling-Wave Tube
Reprinted from: *Sensors* **2023**, *23*, 4762, https://doi.org/10.3390/s23104762 . . . . . . . . . . . . . 65

**Yang Dong, Shaomeng Wang, Jingyu Guo, Zhanliang Wang, Huarong Gong, Zhigang Lu, et al.**
An Angular Radial Extended Interaction Amplifier at the W Band
Reprinted from: *Sensors* **2023**, *23*, 3517, https://doi.org/10.3390/s23073517 . . . . . . . . . . . . . 77

**Xing Xu, Jing Lou, Mingxin Gao, Shiyou Wu, Guangyou Fang and Yindong Huang**
Ultrafast Modulation of THz Waves Based on $MoTe_2$-Covered Metasurface
Reprinted from: *Sensors* **2023**, *23*, 1174, https://doi.org/10.3390/s23031174 . . . . . . . . . . . . . 89

**Hoseong Yoo, Jangsun Kim and Yeong Hwan Ahn**
High-Speed THz Time-of-Flight Imaging with Reflective Optics
Reprinted from: *Sensors* **2023**, *23*, 873, https://doi.org/10.3390/s23020873 . . . . . . . . . . . . . 99

*Article*

# THz Polarimetric Imaging of Carbon Fiber-Reinforced Composites Using the Portable Handled Spectral Reflection (PHASR) Scanner

Kuangyi Xu [1], Zachery B. Harris [1], Paul Vahey [2] and M. Hassan Arbab [1,*]

[1] Department of Biomedical Engineering, Stony Brook University, Stony Brook, NY 11794, USA; kuangyi.xu@stonybrook.edu (K.X.); zachery.harris@stonybrook.edu (Z.B.H.)
[2] Boeing Research & Technology, Seattle, WA 98108, USA; paul.g.vahey@boeing.com
* Correspondence: hassan.arbab@stonybrook.edu

**Abstract:** Recent advancements in novel fiber-coupled and portable terahertz (THz) spectroscopic imaging technology have accelerated applications in nondestructive testing (NDT). Although the polarization information of THz waves can play a critical role in material characterization, there are few demonstrations of polarization-resolved THz imaging as an NDT modality due to the deficiency of such polarimetric imaging devices. In this paper, we have inspected industrial carbon fiber composites using a portable and handheld imaging scanner in which the THz polarizations of two orthogonal channels are simultaneously captured by two photoconductive antennas. We observed significant polarimetric differences between the two-channel images of the same sample and the resulting THz Stokes vectors, which are attributed to the anisotropic conductivity of carbon fiber composites. Using both polarimetric channels, we can visualize the superficial and underlying interfaces of the first laminate. These results pave the way for the future applications of THz polarimetry to the assessment of coatings or surface quality on carbon fiber-reinforced substrates.

**Keywords:** polarimetry; terahertz imaging; non-destructive testing; carbon fiber panels; interwoven carbon fiber; polarimetric PHASR scanner

## 1. Introduction

Recently, terahertz (THz) technologies have rapidly advanced to achieve faster imaging speeds, compact and handheld form factors, and increased signal to noise ratios [1]. Thus, there is a renewed motivation to pursue compelling industrial applications of the THz technology. Due to its unique and distinct sensing capabilities, THz radiation is considered as a promising candidate for nondestructive testing (NDT) in industrial quality control [2]. Imaging with THz pulses (i.e., THz-TDS) can be formed similarly to how Pulse-Echo Ultrasonic Testing is typically applied; it also renders the cross-sectional or depth-resolved pictures of the specimen, which is crucial for evaluating samples of multi-layered structures [3]. This contributes to successful demonstrations made on composite materials, paints on car or airplane bodies, coatings of pharmaceutical tablets, art paintings, etc. [4]. Meanwhile, the implementations of THz technology have expanded beyond benchtop settings, for example, in clinical studies, or on the manufacturing lines of polymer, paper, or pharmaceutical industry [5,6].

Several comprehensive surveys on the NDT applications of THz technologies can be found in recent reviews [7–10]. Terahertz time-domain [11–17] and continuous wave spectroscopy [18–22] have been used in various NDT applications, in part due to the non-ionizing nature [23] and the sub-millimeter resolution of THz waves [24]. These applications include defect and delamination detection [25–28] as well as the measurement of coating thickness [29–31]. The accuracy of these measurements is highly dependent on

**Citation:** Xu, K.; Harris, Z.B.; Vahey, P.; Arbab, M.H. THz Polarimetric Imaging of Carbon Fiber-Reinforced Composites Using the Portable Handled Spectral Reflection (PHASR) Scanner. *Sensors* **2024**, *24*, 7467. https://doi.org/10.3390/s24237467

Academic Editor: Francesco De Leonardis

Received: 28 August 2024
Revised: 11 November 2024
Accepted: 18 November 2024
Published: 22 November 2024

**Copyright:** © 2024 by the authors. Licensee MDPI, Basel, Switzerland. This article is an open access article distributed under the terms and conditions of the Creative Commons Attribution (CC BY) license (https://creativecommons.org/licenses/by/4.0/).

the signal processing techniques employed (e.g., sparse deconvolution [30,32], stationary wavelet denoising [33,34], etc.) to extract the necessary quality control information.

Because most samples feature both intrinsic spatial variations and defect discontinuities, an imaging modality is usually preferred in NDT applications. While many choices of THz imaging systems are available, ranging from a point-scanning device to a single-shot camera, it is challenging to balance between competing performance measures, including the SNR, acquisition speed, resolution, field of view (FOV), and imaging contrast. Such optimizations can be seen in the recent development of our Portable Handheld Spectral Reflection (PHASR) scanners [35], a THz pulsed imaging device based on telecentric raster-scanning using an $f$-$\theta$ objective lens. So far, we have applied this device for the in vivo diagnosis of burn injuries [36–38], assessment of corneal edema [39,40], statistical analysis of speckle patterns [41], chemical imaging through scattering cloaks [42], etc.

In general, after a relatively complex signal processing process implemented on THz-TDS traces, the initial images of THz signals can be converted into maps of complex refractive indices. However, this procedure is less feasible due to poor SNR or limited knowledge of sample properties. Different processing methods are thus explored to interpret the spectroscopic information. Likewise, polarization-sensitive measurements are also incorporated to shed additional insights into sample structures. In this regard, the applications recognized by the existing THz research literature mainly involve (1) the visualization of surface properties (or roughness) [41,43], (2) intrinsic or stress-induced anisotropy of materials [44,45], (3) ellipsometry measurements on thin layers [46], (4) optical activity of biological structures [47,48], and (5) image enhancement [49]. Still, the progress towards the utilization of full polarimetric THz information in sample characterization has been limited. This is partially due to the deficiency of THz polarizing components, and the propagation of uncertainties when individual THz amplitude (intensity) signals are converted into polarimetric parameters [50]. To increase the speed, bandwidth, and accuracy of broadband THz polarization measurements, we have developed a spinning E-O sampling technique [51]. However, due to the need for image formation at high speeds in a portable form factor, using free-space 800 nm amplified laser beams is not practical. Therefore, to enable NDT applications, we recently developed a polarization-sensitive version of our PHASR scanner, enabled by two orthogonal PCA channels. The performance metrics of this device and corresponding calibration methods have been described elsewhere [52].

In this paper, we use the polarization-sensitive PHASR scanner for the imaging of carbon fiber-reinforced polymers (CFRPs) by mapping the THz Stokes vectors. In the NDT analysis of CFRP, the primary problem of interest lies in the evaluation of critical defect types, such as porosity, delamination, fiber defects, and low-velocity impact damage. Early investigations [25,53] had previously concluded that THz imaging "is generally not expected to be a major testing modality (for carbon materials) due to the conductivity of the material and therefore the rather limited penetration of terahertz into composites." [54,55] While not optimistic about that "primary problem", THz imaging can still serve to evaluate coatings on the carbon fiber substrate or characterize superficial properties. The most influential factor in this application of THz imaging is the anisotropic conductivity of CFRP, an intrinsic property that is also responsible for DC electrical or RF shielding testing [56,57]. It seems straightforward that currents along carbon fibers contribute to the reflection and screening of THz waves; however, the characterizations of these spectroscopic properties are limited and often inconsistent. For example, the maximum number of plies that can be resolved in independent THz studies have been significantly different, ranging from as many as eleven or as few as one [58–61]. Our experiments show that, in some CFRP materials provided by Boeing Company (Seattle, WA, USA), the penetration depth for the TE polarization of THz waves is limited to the first ply, whereas most of the TM polarization has been screened at the sample surface. The differences between the reflection spectra of TE and TM polarization are, thus, affected by the magnitude of surface echoes and the presence of subsurface echoes.

## 2. Materials and Methods

Our PHASR scanner incorporates the TERA ASOPS (Asynchronous Optical Sampling) dual-fiber-laser THz spectrometer (Menlo Systems, Inc., Newton, NJ, USA) into a handheld, collocated, telecentric imaging system. The detailed design, polarimetric calibration, and characterization of the performance metrics of the PHASR scanner have been previously reported [52]. Briefly, a THz beam generated by the photoconductive antenna (PCA) in the emitter is collimated using a TPX lens with a 50 mm focal length. The collimated beam is directed towards a gimballed mirror using a high-resistivity silicon beam splitter. The gimballed stage is a two-axis motorized system composed of a goniometer and a rotational stage, which raster-scans the collimated beam over the aperture of a custom-made telecentric $f$-$\theta$ lens. Therefore, the focused beam is always normally incident onto the target and has a constant focal spot size. A free-standing wire grid acts as a polarizing beam splitter, which separates the reflected radiation into the two orthogonal components denoted by X and Y. The signals from the two PCA detectors are converted by two transimpedance amplifiers, and then collected with two digital acquisition (DAQ) cards at synchronized times.

The CFRP testing panels were provided by Boeing Company and include two categories. The first panel is shown in Figure 1, which has multiple layers of coating on a CFRP substrate. The microscopic view of the uncoated substrate, shown in Figure 1b, shows that carbon fibers are aligned in the surface plane, i.e., the ply is unidirectional. The second panel is shown in Figure 2a, whose front surface is fully covered by coating, while the back surface is uncoated. Figure 2b is taken from an uncoated region of the back surface, which is investigated with THz imaging afterwards. This pattern of different "tiles" containing the orthogonal orientations of fibers is due to the plain weave, the dimensions of which are denoted as warp and weft. We will refer to the second class of targets as interwoven CFRP. When fibers aligned to a certain laboratory axis are assigned to a region of the sample, the linear polarization perpendicular to this axis is denoted as TE mode, and the one parallel to this axis is denoted as TM mode. Moving from the warp to the weft of interwoven CFRP, the directions of TE and TM modes will be switched.

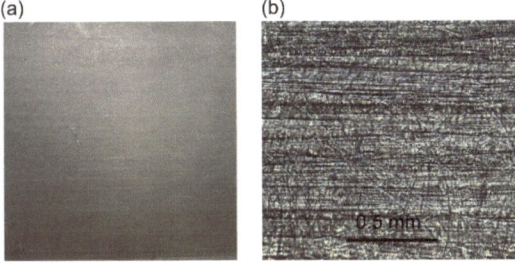

**Figure 1.** (**a**) Front surface of the first test panel from Boeing Company. (**b**) Microscopic image (10×) of the bare substrate, appearing as unidirectional CFRP.

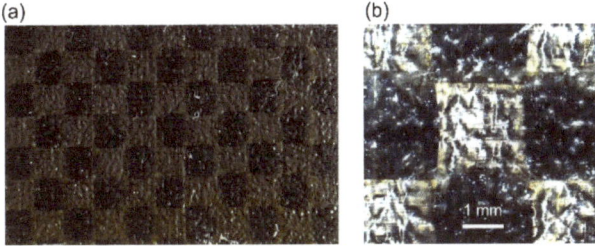

**Figure 2.** (**a**) Back surface of the second test panel from Boeing Company. (**b**) Microscopic image (2.5×) of the back surface, appearing as interwoven (plain-weaved) CFRP.

These composites are placed at the focal plane of the PHASR scanner to form the THz polarimetric images. Each image pixel is recorded in 1 s by averaging 100 replicate traces in the time-domain. The size of FOV and pixel are optimized so that sufficient details of the sample profiles can be captured in reasonable imaging time. The polarization states of the THz source can be manipulated by rotating the PCA emitter and optionally adding a polarizing component. Finally, we have developed wavelet-domain signal analysis tools to investigate and mitigate any surface roughness present on the sample [62–64].

## 3. Results

### 3.1. Polarization-Sensitive Point Measurements

We start the signal processing with individual THz traces to determine the steps in data analysis. The unidirectional CFRP is a highly homogeneous ply and therefore, it is reasonable to compare the sample property as a function of the sample orientation angle. Figure 3a presents the reflected THz time-domain signals at the orientations of 0° and 90°, which is the angle between the fiber axis and the fixed linear polarization of the THz source. After deconvolving by a mirror reference measurement, the resulting spectra of the reflection are presented in Figure 3b. The reflection of the TE mode is overall weaker than that of the TM mode in either Figure 3a or Figure 3b, which is consistent with previous reports [53]. However, there are also exclusive dips below 0.4 THz in the spectra of the TE mode, which seem to be related to the "subsurface echoes" marked by the rectangle in Figure 3a. To substantiate this point, we can turn to the estimated impulse response of the sample, $\hat{h}(t)$, which is obtained from the following relations:

$$w(t) = f_{HF} \exp\left[-(tf_{HF})^2\right] - f_{LF} \exp\left[-(tf_{LF})^2\right], \quad (1)$$

$$\hat{h}(t) = FFT^{-1}\{H(\omega)FFT[w(t)]\}. \quad (2)$$

where $H(\omega)$, the transfer function, is given by

$$H(\omega) = \frac{FFT[sample(t)]}{FFT[reference(t)]}, \quad (3)$$

and $w(t)$ is a double Gaussian function given by the choice of $f_{HF}$ and $f_{LF}$, which have been set to 1.5 THz and 0.3 THz, respectively. This filter was first designed for analyzing the time-domain signals of tissue [65], and has been applied to CFRP later [60]. As shown in Figure 3c, two separate pulses can be recognized in the impulse response of unidirectional CFRP, with a relative time-delay of $\Delta t$ = 7.07 ps. Due to the destructive interference of these two pulses, spectral dips are expected to occur at 0.21 THz and 0.35 THz, which agrees well with the TE mode results in Figure 3b. This pattern of thin-film interference reveals the penetration depth of the THz waves in the testing panel. The TE mode can effectively propagate in the first ply of the unidirectional CFRP and be reflected at the next boundary, whereas the TM mode is significantly screened. For reference, the geometric thickness ($d$) of a single ply in this sample coupon was around 0.2 mm, while its optical thickness given by $\Delta t$ is 1.06 mm, equal to $d$ multiplied by the THz refractive index (which we estimate between 4.4 and 5 [59]).

Next, we apply the same procedure to the measurements of interwoven CFRP, whose impulse responses are presented in Figure 3d. To address the spatial variation, imaging scans of 39 × 39 pixels are conducted on a stationary sample with an interval of 0.25 mm, while the THz emitter is set at an intermediate angle so that the reflections of two orthogonal polarizations are observable simultaneously. The blue and red curves in Figure 3d are captured at different pixels in the same polarimetric image (i.e., in the same detection channel), representing the THz signals of the warp and weft, respectively. Similarly, two separate peaks can be recognized in the TE mode (i.e., the weft); However, the relative time-delay is 3.66 ps, reduced by half compared to the unidirectional sample. It is likely that the TE mode is reflected at the interface between warp and weft within the first ply, and thus, the propagated distance is half of the ply thickness.

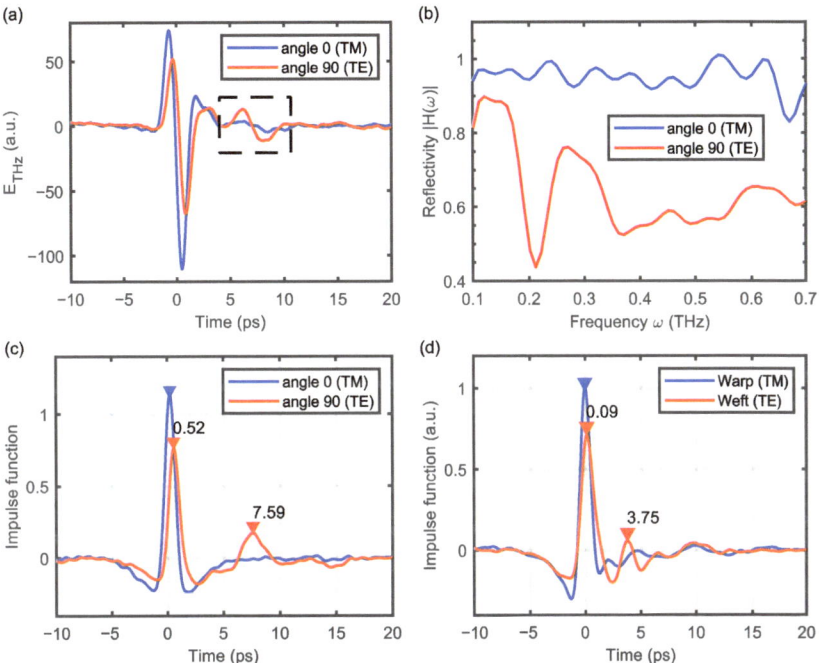

**Figure 3.** (**a**) The THz signals reflected from unidirectional CFRP at the sample orientations of 0° and 90°, corresponding to the TM and TE modes of polarization, respectively. The dashed box shows the difference in the propagation of the TE and TM modes in a single CFRP ply. (**b**,**c**) are the spectra of reflectivity and impulse responses retrieved from signals in (**a**). (**d**) The impulse responses are measured at different locations of interwoven CFRP, where the fiber orientations are different.

### 3.2. Polarization-Sensitive Imaging

A complete set of data includes one pair of polarimetric images from the CFRP sample and another pair obtained using a reference mirror. Thus, the earlier signal processing steps are performed in each pixel. For comparison, we also conducted imaging scans on the unidirectional sample using 41 × 41 pixels with an interval of 0.5 mm. Figure 4 shows the cross-section view of the two CFRP samples, similar to the ultrasonic B-scan. In other words, each column of the image is the impulse response in one pixel, and the entire image constitutes a line scan. Two boundaries can be recognized in these figures: one is around $t = 0$, corresponding to the air–CFRP interface; the other corresponds to the subsurface echoes reflected by the interfaces in the first ply. In Figure 4b,c, the second echoes are more visible in the columns with weaker first echoes, offering an intuitive depiction of the screening effect of carbon fibers on THz E-field. The spatial pattern in the second echoes is reversed when comparing the two polarimetric channels, suggesting that the THz property of CFRP is well described by assuming "binary" responses to the TM and TE modes.

Alternatively, we can visualize the slices in the vicinity of the first and second boundary, resembling the ultrasonic C-scan. Figure 5 shows these *en face* THz images in comparison to the digital photo of a similar sample area. Despite the poorer spatial resolution, the THz images still render the profile of CFRP with abundant details. The excellent complementary relation between the two depth-resolved images from the x and y channels in the left and right column, Figure 5b,c, suggests the binary effect of warp and weft. Also, it can be seen that the spatial variation is related to the degree of screening.

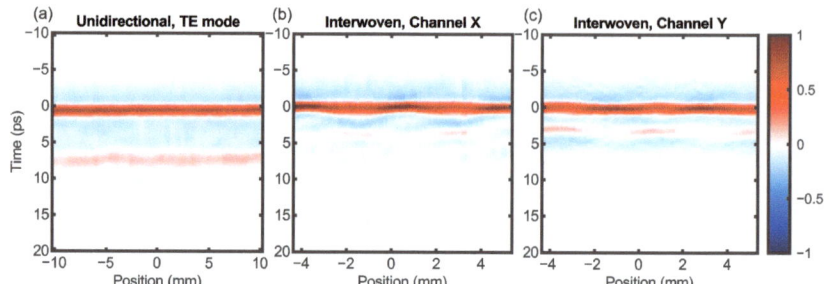

**Figure 4.** Cross-section images (B-scan) of (**a**) the unidirectional CFRP and the interwoven CFRP in the (**b**) X and (**c**) Y channels. The colors are on the same scale and have been extended to [−1, 1].

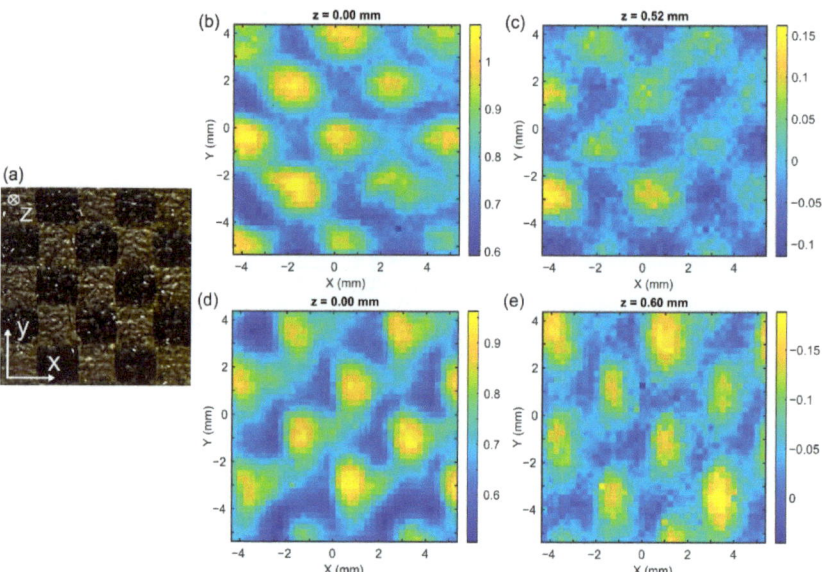

**Figure 5.** (**a**) Photo of the interwoven CFRP, top view. (**b**,**c**) are the C-scanned THz images of the X channel, at the optical depths of z = 0 and z = 0.52 mm, respectively. (**d**,**e**) are the correlated images of the Y channel, at the optical depths of z = 0 and z = 0.60 mm, respectively.

Lastly, we also investigated the joint functions of the two channels, such as the Stokes parameters. Although interesting textures have been observed from these polarimetric contrasts, as shown in Figure 6, it remains challenging to make objective interpretations. Also, according to our discussion on the thin-film interference, any spectral image below 0.4 THz is contributed by both echoes, resembling a blended picture of Figure 5b,c. Therefore, depending on the objectives of the NDT application, signal processing steps may be used to either employ or mitigate the second echo in this specific analysis. At any rate, Figure 6 clearly shows that the complementary regions of the sample are highlighted using the first two elements of the Stokes vectors (i.e., I and Q). Moreover, while the centers of the carbon fiber tiles are shown by the two intensity elements, the edges of the weave patterns are clearly resolved by the other two elements of the Stokes vector incorporating the phase information (i.e., U and V).

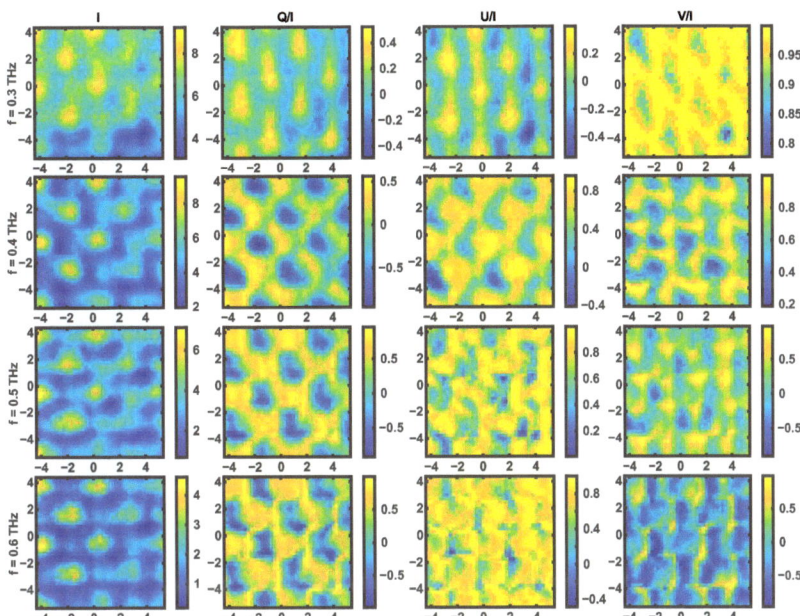

**Figure 6.** The spatial variation in the Stokes parameters *I*, *Q*, *U*, and *V* for the interwoven CFRP at different frequencies. *I* is in arbitrary units while the other Stokes parameters are normalized by *I*.

## 4. Discussion

Our experimental data indicate that the spectral range between 0.4 and 0.6 THz may be ideally suited to investigating carbon fiber samples. However, this ideal spectral range can be heavily affected by the thickness of the carbon fibers. To illustrate this point, Figure 7 shows the sensitivity of THz measurements in the identification of the fiber orientations.

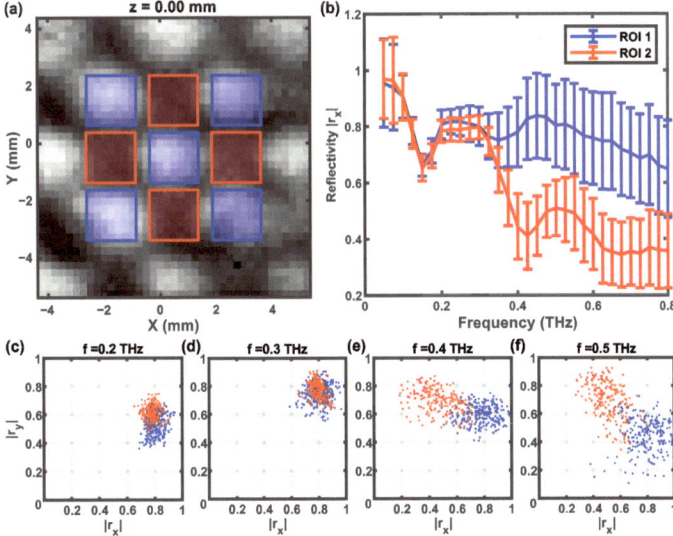

**Figure 7.** (**a**) Two ROIs (blue and red) are selected in the C-scan images of interwoven CFRP. (**b**) The mean value and standard deviation of the reflectivity in the two ROIs. (**c**–**f**) The distribution of pixels in the 2D plane of $|R_x|$ and $|R_y|$, indicating the separation of orthogonal fibers in the 0.4–0.6 THz range.

Figure 7b shows the mean and standard deviations of the reflectivity over two ROIs associated with the warp and weft regions, defined in Figure 7a. The frequency ranges where ROI 1 and ROI 2 are statistically different ($p < 0.01$) have been marked with a gray shade. The contrast between the warp and the weft is more profound in the high-frequency range (>0.4 THz). On the contrary, the contrast at around 0.3 THz is not significant, which can be related to the thin-film interference that occurred in ROI 2. The sensitivity of classification can be seen intuitively in the 2D plane of Rx and Ry, shown in Figure 7c–f, using the reflectivity data of the same pixel in the X and Y channels, respectively.

The polarimetric results presented in this paper can be used to investigate the uniformity and intactness of carbon fiber samples through top coatings that are otherwise opaque to other wavelengths of light. These results also highlight the importance of considering the orientation of the sample and polarization-sensitive imaging using THz ellipsometry or polarimetry instruments. Understanding the strong polarimetric signatures of carbon fiber substrates in various structural forms can influence the interpretation of the recorded data. For instance, if the thickness of various coating layers is to be measured, the polarization of the incident THz beam and the orientation of carbon fibers should be considered.

## 5. Conclusions

We have performed polarization-sensitive THz testing on industrial carbon fiber composites using our PHASR imaging scanner. The spectroscopic properties of these samples are well described by a linear anisotropic model, represented by the axes of TM and TE polarizations. For either the unidirectional or interwoven CFRP material samples, the reflected signals of the TM mode only contain a single pulse, whereas those of the TE mode have a secondary pulse, which is a subsurface echo due to the first ply. The THz spectra of TM and TE polarizations can, thus, provide complementary information about the CFRP, as we have shown by mapping the Stokes vector of the reflected THz beam. This intrinsic anisotropy of CFRP can potentially lead to the application of THz polarimetric imaging in the assessment of coatings or inspection of other issues such as surface quality, fiber intactness, and uniformity.

**Author Contributions:** K.X. collected and processed the data and wrote the original draft; Z.B.H. developed the experimental setup and software; P.V. provided the testing panels and reviewed the paper; M.H.A. supervised the project and edited the paper. All authors have read and agreed to the published version of the manuscript.

**Funding:** This research received no external funding.

**Institutional Review Board Statement:** Not applicable.

**Informed Consent Statement:** Not applicable.

**Data Availability Statement:** The data presented in this study are available upon request from the corresponding author.

**Acknowledgments:** The samples were generously supplied with in-kind support from Boeing Corporation. We thank Bruce R. Davis for the fabrication and preparation of the samples.

**Conflicts of Interest:** The authors declare no conflicts of interest.

## References

1. Bandyopadhyay, A.; Sengupta, A. A Review of the Concept, Applications and Implementation Issues of Terahertz Spectral Imaging Technique. *IETE Tech. Rev.* **2022**, *39*, 471–489. [CrossRef]
2. Wang, B.; Zhong, S.; Lee, T.-L.; Fancey, K.S.; Mi, J. Non-destructive testing and evaluation of composite materials/structures: A state-of-the-art review. *Adv. Mech. Eng.* **2020**, *12*, 1687814020913761. [CrossRef]
3. Dong, J.; Kim, B.; Locquet, A.; McKeon, P.; Declercq, N.; Citrin, D.S. Nondestructive evaluation of forced delamination in glass fiber-reinforced composites by terahertz and ultrasonic waves. *Compos. Part B Eng.* **2015**, *79*, 667–675. [CrossRef]
4. Saeedkia, D. *Handbook of Terahertz Technology for Imaging, Sensing and Communications*; Elsevier: Amsterdam, The Netherlands, 2013.
5. Brinkmann, S.; Vieweg, N.; Gärtner, G.; Plew, P.; Deninger, A. Towards Quality Control in Pharmaceutical Packaging: Screening Folded Boxes for Package Inserts. *J. Infrared Millim. Terahertz Waves* **2017**, *38*, 339–346. [CrossRef]

6. Bauer, M.; Hussung, R.; Matheis, C.; Reichert, H.; Weichenberger, P.; Beck, J.; Matuschczyk, U.; Jonuscheit, J.; Friederich, F. Fast FMCW Terahertz Imaging for In-Process Defect Detection in Press Sleeves for the Paper Industry and Image Evaluation with a Machine Learning Approach. *Sensors* **2021**, *21*, 6569. [CrossRef]
7. Naftaly, M.; Vieweg, N.; Deninger, A. Industrial Applications of Terahertz Sensing: State of Play. *Sensors* **2019**, *19*, 4203. [CrossRef]
8. Ellrich, F.; Bauer, M.; Schreiner, N.; Keil, A.; Pfeiffer, T.; Klier, J.; Weber, S.; Jonuscheit, J.; Friederich, F.; Molter, D. Terahertz Quality Inspection for Automotive and Aviation Industries. *J. Infrared Millim. Terahertz Waves* **2020**, *41*, 470–489. [CrossRef]
9. Tao, Y.H.; Fitzgerald, A.J.; Wallace, V.P. Non-Contact, Non-Destructive Testing in Various Industrial Sectors with Terahertz Technology. *Sensors* **2020**, *20*, 712. [CrossRef]
10. Nsengiyumva, W.; Zhong, S.; Zheng, L.; Liang, W.; Wang, B.; Huang, Y.; Chen, X.; Shen, Y. Sensing and Nondestructive Testing Applications of Terahertz Spectroscopy and Imaging Systems: State-of-the-Art and State-of-the-Practice. *IEEE Trans. Instrum. Meas.* **2023**, *72*, 1–83. [CrossRef]
11. Zhang, D.-D.; Ren, J.-J.; Gu, J.; Li, L.-J.; Zhang, J.-Y.; Xiong, W.-H.; Zhong, Y.-F.; Zhou, T.-Y. Nondestructive testing of bonding defects in multilayered ceramic matrix composites using THz time domain spectroscopy and imaging. *Compos. Struct.* **2020**, *251*, 112624. [CrossRef]
12. Fosodeder, P.; Hubmer, S.; Ploier, A.; Ramlau, R.; van Frank, S.; Rankl, C. Phase-contrast THz-CT for non-destructive testing. *Opt. Express* **2021**, *29*, 15711–15723. [CrossRef] [PubMed]
13. Zhu, P.; Zhang, H.; Robitaille, F.; Maldague, X. Terahertz time-domain spectroscopy for the inspection of dry fibre preforms. *NDT E Int.* **2024**, *145*, 103133. [CrossRef]
14. Liu, Y.; Hu, Y.; Zhang, J.; Liu, H.; Wan, M. Non-Destructive Testing of a Fiber-Web-Reinforced Polymethacrylimide Foam Sandwich Panel with Terahertz Time-Domain Spectroscopy. *Sensors* **2024**, *24*, 1715. [CrossRef] [PubMed]
15. Tu, W.; Zhong, S.; Zhang, Q.; Huang, Y. Rapid diagnosis of corrosion beneath epoxy protective coating using non-contact THz-TDS technique. *Nondestruct. Test. Eval.* **2024**, *39*, 557–572. [CrossRef]
16. Liu, J.; Kong, X.; Cai, C.S.; Peng, H.; Zhang, J. Internal defect characterization of bridge cables based on Terahertz time-domain spectroscopy and deep learning. *Eng. Struct.* **2024**, *314*, 118313. [CrossRef]
17. Liu, Q.; Wang, Q.; Guo, J.; Liu, W.; Xia, R.; Yu, J.; Wang, X. A Transformer-based neural network for automatic delamination characterization of quartz fiber-reinforced polymer curved structure using improved THz-TDS. *Compos. Struct.* **2024**, *343*, 118272. [CrossRef]
18. Zhang, X.; Chang, T.; Wang, Z.; Cui, H.L. Three-Dimensional Terahertz Continuous Wave Imaging Radar for Nondestructive Testing. *IEEE Access* **2020**, *8*, 144259–144276. [CrossRef]
19. Cristofani, E.; Friederich, F.; Wohnsiedler, S.; Matheis, C.; Jonuscheit, J.; Vandewal, M.; Beigang, R. Nondestructive testing potential evaluation of a terahertz frequency-modulated continuous-wave imager for composite materials inspection. *Opt. Eng.* **2014**, *53*, 031211. [CrossRef]
20. Liang, B.; Wang, T.; Shen, S.; Hao, C.; Liu, D.; Liu, J.; Wang, K.; Yang, Z. Occlusion Removal in Terahertz Frequency-Modulated Continuous Wave Non-Destructive Testing Based on Inpainting. *IEEE Trans. Terahertz Sci. Technol.* **2024**, *14*, 699–707. [CrossRef]
21. Moffa, C.; Merola, C.; Magboo, F.J.P.; Chiadroni, E.; Giuliani, L.; Curcio, A.; Palumbo, L.; Felici, A.C.; Petrarca, M. Pigments, minerals, and copper-corrosion products: Terahertz continuous wave (THz-CW) spectroscopic characterization of antlerite and atacamite. *J. Cult. Herit.* **2024**, *66*, 483–490. [CrossRef]
22. Moffa, C.; Curcio, A.; Merola, C.; Migliorati, M.; Palumbo, L.; Felici, A.C.; Petrarca, M. Discrimination of natural and synthetic forms of azurite: An innovative approach based on high-resolution terahertz continuous wave (THz-CW) spectroscopy for Cultural Heritage. *Dye. Pigment.* **2024**, *229*, 112287. [CrossRef]
23. Wallace, V.P.; MacPherson, E.; Zeitler, J.A.; Reid, C. Three-dimensional imaging of optically opaque materials using nonionizing terahertz radiation. *J. Opt. Soc. Am. A* **2008**, *25*, 3120–3133. [CrossRef] [PubMed]
24. Krimi, S.; Klier, J.; Ellrich, F.; Jonuscheit, J.; Urbansky, R.; Beigang, R.; von Freymann, G. An Evolutionary Algorithm Based Approach to Improve the Limits of Minimum Thickness Measurements of Multilayered Automotive Paints. In Proceedings of the 40th International Conference on Infrared, Millimeter, and Terahertz waves (IRMMW-THz), Hong Kong, China, 23–28 August 2015.
25. Ospald, F.; Zouaghi, W.; Beigang, R.; Matheis, C.; Jonuscheit, J.; Recur, B.; Guillet, J.-P.; Mounaix, P.; Vleugels, W.; Bosom, P.; et al. Aeronautics composite material inspection with a terahertz time-domain spectroscopy system. *Opt. Eng.* **2013**, *53*, 031208. [CrossRef]
26. Palka, N.; Panowicz, R.; Chalimoniuk, M.; Beigang, R. Non-destructive evaluation of puncture region in polyethylene composite by terahertz and X-ray radiation. *Compos. Part B Eng.* **2016**, *92*, 315–325. [CrossRef]
27. Li, J.; Yang, L.; He, Y.; Li, W.; Wu, C. Terahertz Nondestructive Testing Method of Oil-paper Insulation Debonding and Foreign Matter Defects. *IEEE Trans. Dielectr. Electr. Insul.* **2021**, *28*, 1901–1908. [CrossRef]
28. Dong, J.; Bianca Jackson, J.; Melis, M.; Giovanacci, D.; Walker, G.C.; Locquet, A.; Bowen, J.W.; Citrin, D.S. Terahertz frequency-wavelet domain deconvolution for stratigraphic and subsurface investigation of art painting. *Opt. Express* **2016**, *24*, 26972–26985. [CrossRef]
29. Shi, H.; Calvelli, S.; Zhai, M.; Ricci, M.; Laureti, S.; Singh, P.; Fu, H.; Locquet, A.; Citrin, D.S. Terahertz Nondestructive Characterization of Conformal Coatings for Microelectronics Packaging. *IEEE Trans. Compon. Packag. Manuf. Technol.* **2024**, *14*, 3–9. [CrossRef]

30. Zhai, M.; Locquet, A.; Citrin, D.S. Terahertz nondestructive layer thickness measurement and delamination characterization of GFRP laminates. *NDT E Int.* **2024**, *146*, 103170. [CrossRef]
31. Zhai, M.; Locquet, A.; Citrin, D.S. Pulsed THz imaging for thickness characterization of plastic sheets. *NDT E Int.* **2020**, *116*, 102338. [CrossRef]
32. Dong, J.L.; Wu, X.L.; Locquet, A.; Citrin, D.S. Terahertz Superresolution Stratigraphic Characterization of Multilayered Structures Using Sparse Deconvolution. *IEEE Trans. Terahertz Sci. Technol.* **2017**, *7*, 260–267. [CrossRef]
33. Dong, J.; Locquet, A.; Citrin, D.S. Terahertz Quantitative Nondestructive Evaluation of Failure Modes in Polymer-Coated Steel. *IEEE J. Sel. Top. Quantum Electron.* **2017**, *23*, 1–7. [CrossRef]
34. Zhai, M.; Locquet, A.; Roquelet, C.; Ronqueti, L.A.; Citrin, D.S. Thickness characterization of multi-layer coated steel by terahertz time-of-flight tomography. *NDT E Int.* **2020**, *116*, 102358. [CrossRef]
35. Harris, Z.B.; Arbab, M.H. Terahertz PHASR Scanner With 2 kHz, 100 ps Time-Domain Trace Acquisition Rate and an Extended Field-of-View Based on a Heliostat Design. *IEEE Trans. Terahertz Sci. Technol.* **2022**, *12*, 619–632. [CrossRef] [PubMed]
36. Khani, M.E.; Osman, O.B.; Harris, Z.B.; Chen, A.; Zhou, J.W.; Singer, A.J.; Arbab, M.H. Accurate and early prediction of the wound healing outcome of burn injuries using the wavelet Shannon entropy of terahertz time-domain waveforms. *J. Biomed. Opt.* **2022**, *27*, 116001. [CrossRef]
37. Osman, O.B.; Harris, Z.B.; Khani, M.E.; Zhou, J.W.; Chen, A.; Singer, A.J.; Arbab, M.H. Deep neural network classification of in vivo burn injuries with different etiologies using terahertz time-domain spectral imaging. *Biomed. Opt. Express* **2022**, *13*, 1855–1868. [CrossRef]
38. Khani, M.E.; Harris, Z.B.; Osman, O.B.; Singer, A.J.; Arbab, M.H. Triage of in vivo burn injuries and prediction of wound healing outcome using neural networks and modeling of the terahertz permittivity based on the double Debye dielectric parameters. *Biomed. Opt. Express* **2023**, *14*, 918–931. [CrossRef]
39. Chen, A.; Harris, Z.B.; Virk, A.; Abazari, A.; Varadaraj, K.; Honkanen, R.; Arbab, M.H. Assessing Corneal Endothelial Damage Using Terahertz Time-Domain Spectroscopy and Support Vector Machines. *Sensors* **2022**, *22*, 9071. [CrossRef]
40. Virk, A.S.; Harris, Z.B.; Arbab, M.H. Development of a terahertz time-domain scanner for topographic imaging of spherical targets. *Opt. Lett.* **2021**, *46*, 1065–1068. [CrossRef]
41. Xu, K.; Harris, Z.B.; Arbab, M.H. Polarimetric imaging of back-scattered terahertz speckle fields using a portable scanner. *Opt. Express* **2023**, *31*, 11308–11319. [CrossRef]
42. Khani, M.E.; Harris, Z.B.; Liu, M.; Arbab, M.H. Multiresolution spectrally-encoded terahertz reflection imaging through a highly diffusive cloak. *Opt. Express* **2022**, *30*, 31550–31566. [CrossRef]
43. Wan, M.; Yuan, H.; Healy, J.J.; Sheridan, J.T. Terahertz confocal imaging: Polarization and sectioning characteristics. *Opt. Lasers Eng.* **2020**, *134*, 106182. [CrossRef]
44. Okano, M.; Watanabe, S. Internal Status of Visibly Opaque Black Rubbers Investigated by Terahertz Polarization Spectroscopy: Fundamentals and Applications. *Polymers* **2018**, *11*, 9. [CrossRef] [PubMed]
45. Deng, Y.; McKinney, J.A.; George, D.K.; Niessen, K.A.; Sharma, A.; Markelz, A.G. Near-Field Stationary Sample Terahertz Spectroscopic Polarimetry for Biomolecular Structural Dynamics Determination. *ACS Photonics* **2021**, *8*, 658–668. [CrossRef]
46. Chen, X.; Pickwell-MacPherson, E. An introduction to terahertz time-domain spectroscopic ellipsometry. *APL Photonics* **2022**, *7*, 071101. [CrossRef]
47. Hayut, I.; Ben Ishai, P.; Agranat, A.J.; Feldman, Y. Circular polarization induced by the three-dimensional chiral structure of human sweat ducts. *Phys. Rev. E* **2014**, *89*, 042715. [CrossRef]
48. Choi, W.J.; Cheng, G.; Huang, Z.; Zhang, S.; Norris, T.B.; Kotov, N.A. Terahertz circular dichroism spectroscopy of biomaterials enabled by kirigami polarization modulators. *Nat. Mater.* **2019**, *18*, 820–826. [CrossRef]
49. Zhang, Y.; Wang, C.; Huai, B.; Wang, S.; Zhang, Y.; Wang, D.; Rong, L.; Zheng, Y. Continuous-Wave THz Imaging for Biomedical Samples. *Appl. Sci.* **2021**, *11*, 71. [CrossRef]
50. Naftaly, M. *Terahertz Metrology*; Artech House: London, UK, 2015.
51. Xu, K.; Liu, M.; Arbab, M.H. Broadband terahertz time-domain polarimetry based on air plasma filament emissions and spinning electro-optic sampling in GaP. *Appl. Phys. Lett.* **2022**, *120*, 181107. [CrossRef]
52. Harris, Z.B.; Xu, K.; Arbab, M.H. A handheld polarimetric imaging device and calibration technique for accurate mapping of terahertz Stokes vectors. *Sci. Rep.* **2024**, *14*, 17714. [CrossRef]
53. Karpowicz, N.; Dawes, D.; Perry, M.J.; Zhang, X.-C. Fire damage on carbon fiber materials characterized by THz waves. *Proc. SPIE* **2006**, *6212*, 62120G.
54. Howell, P.A. *Nondestructive Evaluation (NDE) Methods and Capabilities Handbook*; TM−2020-220568; Langley Research Center, NASA: Hampton, VA, USA, 2020.
55. Ibrahim, M.E. Nondestructive evaluation of thick-section composites and sandwich structures: A review. *Compos. Part A Appl. Sci. Manuf.* **2014**, *64*, 36–48. [CrossRef]
56. Zhao, Q.; Zhang, K.; Zhu, S.; Xu, H.; Cao, D.; Zhao, L.; Zhang, R.; Yin, W. Review on the Electrical Resistance/Conductivity of Carbon Fiber Reinforced Polymer. *Appl. Sci.* **2019**, *9*, 2390. [CrossRef]
57. Angulo, L.M.D.; Francisco, P.G.d.; Gallardo, B.P.; Martinez, D.P.; Cabello, M.R.; Bocanegra, D.E.; Garcia, S.G. Modeling and Measuring the Shielding Effectiveness of Carbon Fiber Composites. *IEEE J. Multiscale Multiphysics Comput. Tech.* **2019**, *4*, 207–213. [CrossRef]

58. Im, K.-H.; Hsu, D.K.; Chiou, C.-P.T.; Barnard, D.J.; Kim, S.-K.; Kang, S.-J.; Cho, Y.T.; Jung, J.-A.; Yang, I.Y. Influence of Terahertz Waves on Unidirectional Carbon Fibers in CFRP Composite Materials. *Mater. Sci.* **2014**, *20*, 457–463. [CrossRef]
59. Zhang, J.; Shi, C.; Ma, Y.; Han, X.; Li, W.; Chang, T.; Wei, D.; Du, C.; Cui, H.-L. Spectroscopic study of terahertz reflection and transmission properties of carbon-fiber-reinforced plastic composites. *Opt. Eng.* **2015**, *54*, 054106. [CrossRef]
60. Sørgård, T.; van Rheenen, A.; Haakestad, M. Terahertz imaging of composite materials in reflection and transmission mode with a time-domain spectroscopy system. *Proc. SPIE* **2016**, *9747*, 974714.
61. Dong, J.; Pomarède, P.; Chehami, L.; Locquet, A.; Meraghni, F.; Declercq, N.F.; Citrin, D.S. Visualization of subsurface damage in woven carbon fiber-reinforced composites using polarization-sensitive terahertz imaging. *NDT E Int.* **2018**, *99*, 72–79. [CrossRef]
62. Khani, M.E.; Arbab, M.H. Translation-Invariant Zero-Phase Wavelet Methods for Feature Extraction in Terahertz Time-Domain Spectroscopy. *Sensors* **2022**, *22*, 2305. [CrossRef]
63. Arbab, M.H.; Winebrenner, D.; Thorsos, E.; Chen, A. Application of wavelet transforms in terahertz spectroscopy of rough surface targets. *Proc. SPIE* **2010**, *7601*, 760106.
64. Arbab, M.H.; Chen, A.; Thorsos, E.; Winebrenner, D.; Zurk, L. Effect of surface scattering on terahertz time domain spectroscopy of chemicals. *Proc. SPIE* **2008**, *6893*, 68930C.
65. Ruth, M.W.; Bryan, E.C.; Vincent, P.W.; Richard, J.P.; Donald, D.A.; Edmund, H.L.; Michael, P. Terahertz pulse imaging in reflection geometry of human skin cancer and skin tissue. *Phys. Med. Biol.* **2002**, *47*, 3853.

**Disclaimer/Publisher's Note:** The statements, opinions and data contained in all publications are solely those of the individual author(s) and contributor(s) and not of MDPI and/or the editor(s). MDPI and/or the editor(s) disclaim responsibility for any injury to people or property resulting from any ideas, methods, instructions or products referred to in the content.

*Communication*

# Research of a 0.14 THz Dual-Cavity Parallel Structure Extended Interaction Oscillator

**Chuanhong Xiao [1], Ruizhe Ren [1], Zhenhua Wu [1,*], Yijun Li [2], Qing You [1], Zongjun Shi [1], Kaichun Zhang [1], Xiaoxing Chen [1], Mingzhou Zhan [3], Diwei Liu [1], Renbin Zhong [1] and Shenggang Liu [1]**

[1] School of Electronic Science and Engineering, University of Electronic Science and Technology of China, Chengdu 610054, China; chuanhong_x@163.com (C.X.); 202221020217@std.uestc.edu.cn (R.R.); 202021020527@std.uestc.edu.cn (Q.Y.); shizongjun@uestc.edu.cn (Z.S.); zh.kch@163.com (K.Z.); cxxmarshal@uestc.edu.cn (X.C.); dwliu212220@163.com (D.L.); rbzhong@uestc.edu.cn (R.Z.); liusg@uestc.edu.cn (S.L.)

[2] State Key Laboratory of Polymer Materials Engineering, Polymer Research Institute, Sichuan University, Chengdu 610041, China; yijunliruddph@gmail.com

[3] School of Physics, University of Electronic Science and Technology of China, Chengdu 610054, China; mzzhan@uestc.edu.cn

* Correspondence: wuzhenhua@uestc.edu.cn

**Abstract:** This paper presents a method to enhance extended interaction oscillator (EIO) output power based on a dual-cavity parallel structure (DCPS). This stucture consists of two conventional ladder-line structures in parallel through a connecting structure, which improves the coupling efficiency between the cavities. The dual output power fusion structure employs an H-T type combiner as the output coupler, which can effectively combine the two input waves in phase to further increase the output power. The dispersion characteristics, coupling impedance, and field distribution of the DCPS are investigated through numerical and simulation calculations, and the optimal operating parameters and output structure are obtained by PIC simulation. At an operating voltage of 12.6 kV, current density of 200 A/cm$^2$, and longitudinal magnetic field of 0.5 T, the DCPS EIO exhibits an output power exceeding 600 W at a frequency of 140.6 GHz. This represents a nearly threefold enhancement compared with the 195 W output of the conventional ladder-line EIO structure. These findings demonstrate the significant improvement in output power and interaction efficiency achieved by the DCPS for the EIO device.

**Keywords:** extended interaction oscillator; terahertz radiation; vacuum electronic device; dual cavity parallel; particle simulation

## 1. Introduction

The pursuit of high-frequency electromagnetic waves continues to propel technological advancements across various domains. One particularly promising frontier in this endeavor lies within the terahertz (THz) frequency range, spanning from 0.1 to 10 THz. This region of the electromagnetic spectrum holds immense potential for a diverse array of applications, including communication, imaging, sensing, and spectroscopy. However, the effective utilization of the THz regime poses significant challenges, primarily due to the limited availability of a suitable radiation source [1–3]. In this paper, vacuum electronic devices (VEDs) emerge as an intriguing avenue, capitalizing on the unique characteristics of VEDs to facilitate high-power, high-frequency operations. Among the notable classes of VEDs, extended interaction oscillators (EIOs) stand out for their exceptional attributes, such as high output power, superior frequency stability, and impressive phase noise performance. Some EIO parameters are shown in Table 1 [4–8].

Table 1. Parameters of the current EIOs.

| Institution | Frequency (GHz) | Voltage (kV) | Current (A) | Power (W) |
|---|---|---|---|---|
| CPI | 93.8 | 20.3 | 0.69 | 1400 |
| UESTC | 140 | 18 | 1.5 | 661 |
| CPI | 140 | / | / | 200 |
| UESTC | 220 | 16.6 | 3.2 | 500 |
| CPI | 214.5 | 11 | 0.095 | 13.3 |
| UESTC | 300 | 14.8 | 0.25 | 250 |

However, as the frequency increases from the millimeter-wave to the terahertz band, the structure size of the traditional single-cavity EIO decreases, leading to a decline in power capacity that is challenging to improve. In order to solve the limitation of single-cavity EIO power, the combination of multibeam and multicavity is considered. In this case, this paper proposes a dual-cavity parallel structure of EIO, which connects two traditional ladder-line structure EIOs through a rectangular connection structure, and effectively improves the efficiency of the beam-wave interaction. Due to the symmetry of the dual-cavity parallel structure, both cavities can output electromagnetic waves with the same power and frequency, so the power fusion structure is formed through the integration of waveguides, which greatly improves the output power and break through the power limitations of the traditional single-cavity EIOs [9–20].

## 2. Structural Design and Cold Cavity Analysis

EIO incorporates a slow-wave structure, exemplified by a ladder-line configuration, as a crucial constituent component, which ensures the coupling impedance, power, and efficiency, while realizing the miniaturization of the device, and also ensures that it is able to interact well with the sheet beam.

The characteristic impedance ($R/Q$) is a critical factor in measuring the performance of resonant cavities, which can be calculated as follows:

$$\frac{R}{Q} = \frac{\left( \int_{-\infty}^{\infty} \left| \frac{1}{y} \int_{-\frac{y}{2}}^{\frac{y}{2}} E_z dy \right| dz \right)^2}{2\omega W_s} \tag{1}$$

where $E_z$ represents the electric field strength in the Z-direction, $\omega$ represents the frequency of the resonant cavity, and $W_s$ represents the total stored energy.

As illustrated in Figure 1a, the device features a conventional single-cavity structure. Compared to the $TM_{11}$ mode, the $TM_{31}$ mode allows for a more complete interaction between the sheet beam and the electromagnetic field, thereby enabling the achievement of a high output power. To further enhance the output power of the EIO, a dual-cavity parallel structure is proposed in this paper, as shown in Figure 1b, which connects two conventional ladder-line structure EIOs in parallel through a rectangular connecting structure, and this dual-cavity parallel structure EIO not only retains the mechanism of the interaction between the conventional single-cavity structure EIO and the higher order modes as well as the higher efficiency of the beam-wave interactions, but also effectively enhances the coupling efficiency between the cavities. Through extensive theoretical numerical and simulation calculations to optimize the structural parameters, the operating frequency has been determined to be 0.14 THz. The specific parameter values are shown in Table 2.

The dispersion curve is obtained as shown in Figure 2 after calculation and analysis. The difference between the dispersion characteristics of the two structures in the same operating mode is not significant. The operating frequency of the $TM_{31}$-$2\pi$ mode is significantly higher than that of the $TM_{11}$-$2\pi$ mode, and the operating frequency of the $TM_{31}$-$2\pi$ mode of the DCPS EIO is 141.03 GHz. In the DCPS EIO, the characteristic impedance of the $TM_{31}$-$2\pi$ mode (464 Ω) is much larger than that of the $TM_{11}$-$2\pi$ mode (35 Ω). Therefore, the structure effectively suppresses the $TM_{11}$-$2\pi$ mode, the $TM_{31}$-$2\pi$ mode is more easily stimulated for oscillation, and this DCPS has a larger cavity volume and, therefore, a larger power capacity.

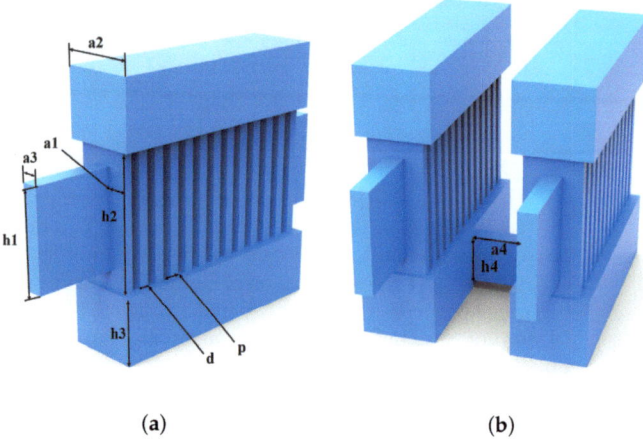

(a)     (b)

**Figure 1.** Physical model of the resonant cavity. (**a**) The single-cavity structure. (**b**) The dual-cavity parallel structure.

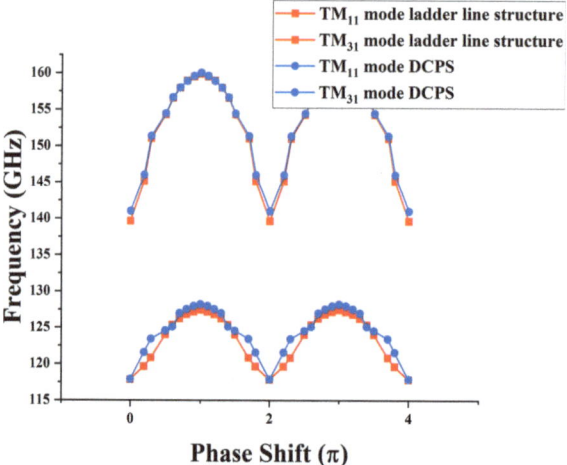

**Figure 2.** The cold cavity characteristics.

**Table 2.** Parameters of the resonant cavity.

| Parameter | Quantity | Value (mm) |
| --- | --- | --- |
| a1 | Gap depth | 0.42 |
| a2 | Coupling cavity width | 1.68 |
| a3 | Beam tunnel width | 0.40 |
| a4 | Coupling cavity width | 1.68 |
| h1 | Beam tunnel height | 2.00 |
| h2 | Gap height | 3.00 |
| h3 | Coupling cavity height | 1.50 |
| h4 | Coupling cavity height | 1.10 |
| d | Gap width | 0.22 |
| p | Period length | 0.44 |

The electric field distribution of the two structures of EIO and the $E_Z$ electric field strengths measured along the two lines are shown in Figure 3. Figure 3a,b depict the electric field profiles of the $TM_{31}$ and $TM_{11}$ modes, respectively, along with the centerlines in the Y and Z directions. The electric field intensity is then calculated along these central axes. Figure 3c shows the amplitude distribution of the $E_Z$ field in the Z-direction. The results show that the $E_Z$ field in the gap of the $TM_{31}$ mode is large and stable, and much larger than that of the $TM_{11}$ mode, which indicates that the $TM_{31}$ mode has a stronger beam-wave interaction capability. Figure 3d shows the amplitude distribution of the $E_Z$ field in the Y-direction. The results show that the electric field strength in the beam tunnel of the $TM_{31}$ mode is larger than that of the $TM_{11}$ mode, which is favorable to the beam-wave interaction, and that the excessive electric field strength of the $TM_{11}$ mode in the coupling cavity at the lower end may lead to the unbalanced coupling of the electromagnetic wave. Through the cold cavity analysis, the $TM_{31}$ mode was finally selected as the operating mode for the DCPS EIO. The $S_{11}$ parameters of the EIO are shown in Figure 4, from which it can be concluded that there is a good scattering parameter at the output frequency point, the power generated can be well coupled away from the output port.

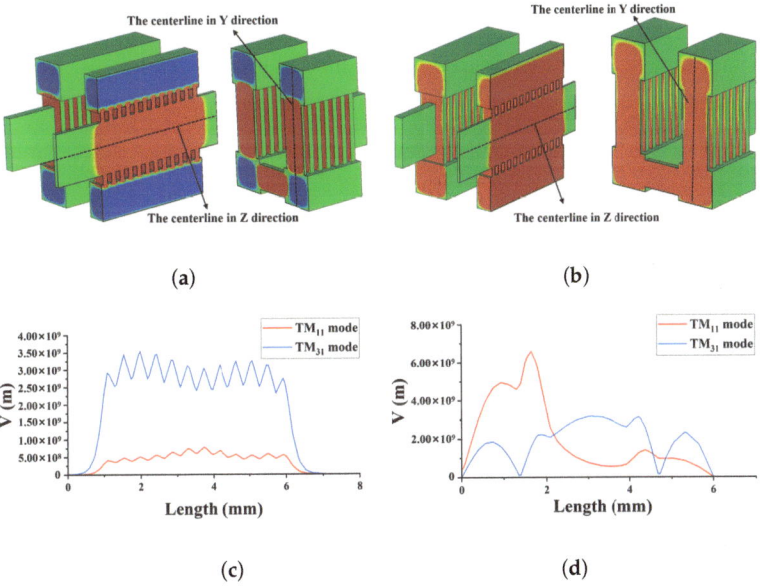

**Figure 3.** $E_Z$ field distribution. (**a**) $E_Z$ field distribution in the $TM_{31}$-$2\pi$ mode. (**b**) $E_Z$ field distribution in the $TM_{11}$-$2\pi$ mode. (**c**) Amplitude of the centerline in the Z-direction. (**d**) Amplitude of the centerline in the Y-direction.

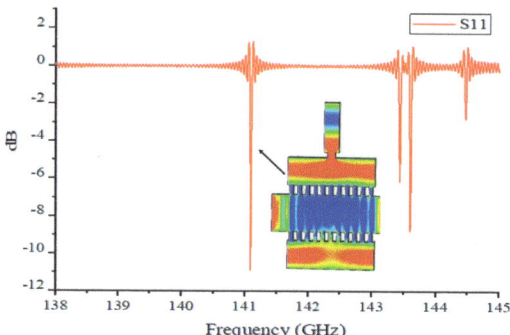

**Figure 4.** $S_{11}$ parameters of the DCPS EIO and the field distribution of the $TM_{31}$ mode.

## 3. PIC Simulation

The optimal structural parameters were determined through cold cavity analysis. To verify the efficacy of the DCPS EIO in enhancing the beam-wave interaction efficiency, PIC simulations were conducted to determine the optimal operating parameters. The size of the electron beam was 2 mm × 0.3 mm, and the current density of each beam was 200 A/cm² with an operating voltage of 12.6 kV and longitudinal magnetic field of 0.5 T. Considering the effect of machining roughness and ohmic loss in the DCPS EIO, the Hammerstad-Bekkadal formula can be used to predict the excessive resistance loss and the formula is shown below:

$$\sigma_{eff} = \sigma_0 \cdot \left\{ 1 + \frac{2}{\pi} \arctan\left[1.4 \left(\frac{h}{\delta}\right)^2\right] \right\}^{-2} \quad (2)$$

where $\sigma$ is the conductivity of the ideal smooth surface ($\sigma_{Cu} = 5.8 \times 10^7$ S/m), $\sigma_{eff}$ is the effective conductivity of the rough surface, $\delta$ is the skin depth under the ideal smooth plane, and h is the root mean square (RMS) height of the surface. The equivalent conductivity was calculated to be $3 \times 10^7$ S/m.

In order to achieve the purpose of significantly increasing the output power, on top of this single output structure, a single output structure with the same structural dimensions was set above both the left and right cavities, and then a 90° curved waveguide was utilized to connect these two single output structures to a rectangular waveguide, which was then connected to the standard waveguide above, to realize the integration of the output structure and to complete the fusion of power. The three structures with output structures are shown in Figure 5.

Figure 6 shows the comparison of the output power of the normal structure, the dual-cavity parallel structure, and the dual-cavity parallel structure with power fusion. Obviously, the output power obtained at 140.6 GHz in the dual-cavity parallel structure was 366 W, which was nearly twice the output power of the normal structure, and both the left and right cavities can output electromagnetic waves, and the output power with power fusion was 610 W. In this study, the coupling of electromagnetic waves was achieved by incorporating a connecting structure between the two cavities. As such, the analysis of the coupling cavity is of critical importance. As shown in Figure 7, the output power results were plotted by changing the length and width of the output cavity. The final dimensions of the coupling cavity were 1.68 mm × 0.44 mm. The phase space of the EIO and the electron beam are shown in Figure 8, where it can be observed that the electron beam was continuously modulated in the resonant cavity and reached the optimal modulation at the end.

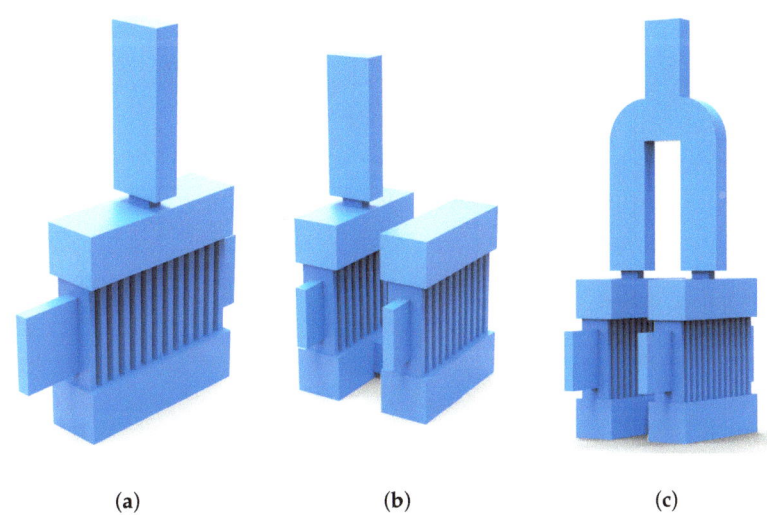

**Figure 5.** Physical model of the resonant cavity with output structure. (**a**) The single-cavity structure. (**b**) The DCPS structure. (**c**) The DCPS structure with power fusion.

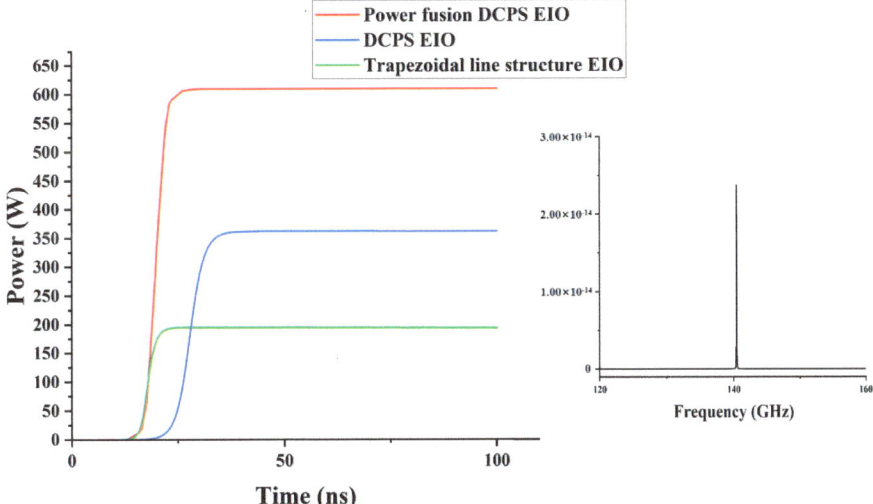

**Figure 6.** The output power of the three structure's EIOs and the output frequency of the DCPS EIO.

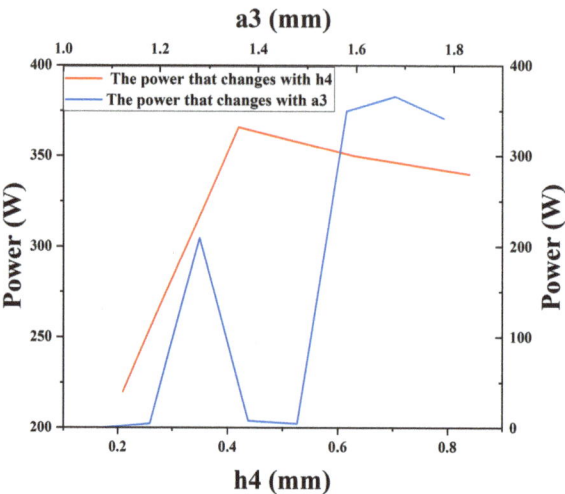

**Figure 7.** Effect with changing the connection structure on the output power.

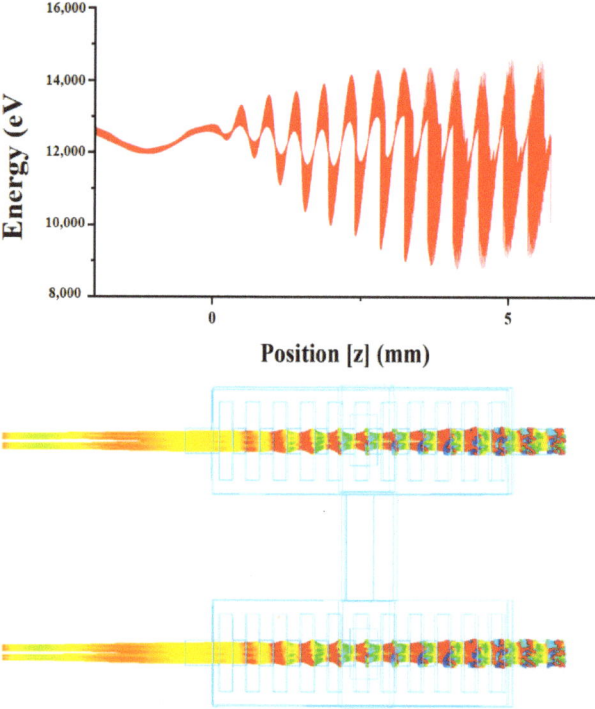

**Figure 8.** Electron beam energy phase space and cluster.

## 4. Discussion

To improve the output power of the EIO operating in the terahertz frequency range, this study proposes a new structure called DCPS (dual-cavity parallel structure). DCPS achieves multi-cavity coupling through a coupling structure at the bottom of the cavities. Numerical and simulation analyses of the DCPS demonstrate several key advantages.

First, by adopting DCPS, the power capacity is significantly improved compared to the traditional ladder-line EIO structure, providing critical support for the realization of high-power terahertz sources. This is a crucial advancement, as high-power terahertz sources are in high demand for many sensing and imaging applications.

Second, the symmetric design of the DCPS cavities allows for convenient integration of a power fusion structure, such as the H-T type combiner used in this study, to further increase the overall output power. Under the operating conditions of 12.6 kV voltage, 200 A/cm$^2$ current density, and 0.5 T longitudinal magnetic field, DCPS EIO exhibits an impressive output power exceeding 600 W at a frequency of 140.6 GHz. This represents nearly a three-fold improvement compared with the 195 W output power of the conventional ladder-line EIO. Compared with other EIOs in the same frequency band, the output power is at an optimal level. The significant improvements in output power and interaction efficiency of DCPS can be attributed to the effective coupling between the two parallel cavities and the power combining structure. The design of the electro-optical system will be carried out in a follow-up study and the structure will be processed using the Ultraviolet Lithographie Galvanoformung Abformung (UV-LIGA) technique. This power fusion approach provides a new idea for the subsequent development of high-power terahertz sources.

High-power terahertz sources have a profound impact on the sensor field. For instance, in security screening, the increased output power enables more sensitive and reliable detection of concealed threats, such as explosives and weapons, through enhanced terahertz imaging technology. This capability is particularly valuable in crowded environments, where rapid and accurate screening can significantly enhance public safety. In the medical domain, high-power terahertz sources can improve the quality and penetration depth of imaging, playing a crucial role in the early detection and diagnosis of cancer. By providing more detailed images of tissue structures, these sources can help clinicians identify abnormalities that may be missed with traditional imaging methods. Additionally, these high-power terahertz sources find applications in industrial sectors, such as non-destructive testing and quality control, helping companies ensure the safety and compliance of their products while reducing waste and downtime.

Therefore, I believe that further research on high-power terahertz sources can significantly facilitate the advancement of sensors. This not only enhances the performance of existing technologies, but also drives innovation in new sensor applications, laying a solid foundation for various fields.

## 5. Conclusions

This paper proposes a dual-cavity parallel structure-based extended interaction oscillator that effectively addresses the problem of low output power in traditional single-cavity EIOs at THz frequencies. Under optimal operating conditions, this EIO device achieves an output power exceeding 600 W at 140.6 GHz, nearly tripling the output of conventional EIOs. These findings demonstrate the immense potential of the DCPS approach in developing high-power THz EIO devices and provide an effective technical pathway for the advancement of high-power THz sources.

**Author Contributions:** Conceptualization, C.X. and Z.W.; methodology, Q.Y. and R.R.; validation, K.Z., D.L. and R.Z.; formal analysis, Q.Y., C.X. and R.R.; investigation, Q.Y. and R.R.; resources, S.L. and Z.W.; data curation, R.R. and Z.W.; writing—original draft preparation, R.R. and Z.W.; writing—review and editing, R.R. and Z.W.; visualization, X.C. and C.X.; supervision, Z.S., S.L. and Y.L.; project administration, S.L. and Z.W.; funding acquisition, M.Z., S.L. and Y.L. All authors have read and agreed to the published version of the manuscript.

**Funding:** This study was supported by the National Key Research and Development Program of China (grant numbers: 2020YFA0714001, 2018YFF01013001 and 2017YFA0701000), the Natural Science Foundation of China (grant numbers: 62471119, 61988102, 61921002 and 62071108), the Key Science and Technology Program of Sichuan Province (grant number: 23ZDYF0471), China Postdoctoral

Science Foundation (2023T160443), and in part by the Natural Science Foundation of Sichuan Province (grant number: 2022NSFSC0513).

**Data Availability Statement:** The data that support the findings of this study are available from the corresponding author upon reasonable request.

**Conflicts of Interest:** The authors declare no conflicts of interest.

# References

1. Booske, J.H.; Dobbs, R.J.; Joye, C.D.; Kory, C.L.; Neil, G.R.; Park, G.S.; Park, J.; Temkin, R.J. Vacuum Electronic High Power Terahertz Sources. *IEEE Trans. Terahertz Sci. Technol.* **2011**, *1*, 54–75. [CrossRef]
2. Siegel, P. Terahertz technology. *IEEE Trans. Microw. Theory Tech.* **2002**, *50*, 910–928. [CrossRef]
3. Zhang, X.; Zhang, R.; Wang, Y. Research on a High-Order Mode Multibeam Extended-Interaction Oscillator With Coaxial Structure. *IEEE Trans. Plasma Sci.* **2020**, *48*, 1902–1909. [CrossRef]
4. Chang, Z.; Meng, L.; Li, H.; Wang, B.; Yuan, X.; Xu, C.; Peng, R.; Yin, Y. A High-Efficiency Dual-Cavity Extended Interaction Oscillator. *IEEE Trans. Electron Devices* **2020**, *67*, 335–340. [CrossRef]
5. Sarwar, M.S.; Niu, X.; Zhang, T.; Liu, Y. Design of a Peculiar $TM_{35}$ Transverse Mode THz Extended Interaction Oscillator for Multibeam kW-Class Operation. *IEEE Trans. Plasma Sci.* **2024**, *52*, 707–714. [CrossRef]
6. Joye, C.D.; Cook, A.M.; Calame, J.P.; Abe, D.K.; Vlasov, A.N.; Chernyavskiy, I.A.; Nguyen, K.T.; Wright, E.L.; Pershing, D.E.; Kimura, T.; et al. Demonstration of a High Power, Wideband 220-GHz Traveling Wave Amplifier Fabricated by UV-LIGA. *IEEE Trans. Electron Devices* **2014**, *61*, 1672–1678. [CrossRef]
7. Li, J.; Wu, Z.; Liu, D.; Wang, W.; Zhao, T.; Zhong, R.; Shi, Z.; Zhang, K.; Duan, Z.; Wei, Y.; et al. Novel 0.22-THz Extended Interaction Oscillator Based on the Four-Sheet-Beam Orthogonal Interconnection Structure. *IEEE Trans. Electron Devices* **2023**, *70*, 1917–1922. [CrossRef]
8. Zu, Y.; Lan, Y.; Yuan, X.; Xu, X.; Chen, Q.; Li, H.; Cole, M.T.; Yin, Y.; Wang, B.; Meng, L.; et al. Research on a Highly Overmoded Slow Wave Circuit for 0.3-THz Extended Interaction Oscillator. *IEEE Trans. Electron Devices* **2023**, *70*, 2165–2169. [CrossRef]
9. Dong, Y.; Wang, S.; Guo, J.; Wang, Z.; Tang, T.; Gong, H.; Lu, Z.; Duan, Z.; Gong, Y. A 0.14 THz Angular Radial Extended Interaction Oscillator. *IEEE Trans. Electron Devices* **2022**, *69*, 1468–1473. [CrossRef]
10. Qing, J.; Niu, X.; Zhang, T.; Liu, Y.; Guo, G.; Li, H. Design and research of a novel structure for extended interaction oscillators. *Phys. Plasmas* **2022**, *29*. [CrossRef]
11. Li, J.; Liu, D.; Ren, R.; Xiao, C.; Shi, Z.; Zhao, T.; Hu, M.; Wei, Y.; Duan, Z.; Gong, Y.; et al. A Novel 2-D Slotted Structure Extended Interaction Oscillator. *IEEE Trans. Electron Devices* **2023**, *70*, 2780–2785. [CrossRef]
12. Qing, J.; Niu, X.; Liu, Y.; Guo, G.; Li, H. Design and Cold Test on the Slow Wave Structure of a Wide-Voltage Tuned and High-Power Extended Interaction Oscillator in W-Band. *IEEE Trans. Plasma Sci.* **2023**, *51*, 381–385. [CrossRef]
13. Xu, C.; Meng, L.; Paoloni, C.; Qin, Y.; Bi, L.; Wang, B.; Li, H.; Yin, Y. A 0.35-THz Extended Interaction Oscillator Based on Overmoded and Bi-Periodic Structure. *IEEE Trans. Electron Devices* **2021**, *68*, 5814–5819. [CrossRef]
14. Qing, J.; Niu, X.; Zhang, T.; Liu, Y.; Guo, G.; Li, H. THz Radiation from a $TM_{51}$ Mode Sheet Beam Extended Interaction Oscillator With Low Injection. *IEEE Trans. Plasma Sci.* **2022**, *50*, 1081–1086. [CrossRef]
15. Wang, J.; Wan, Y.; Xu, D.; Li, X.; Dai, Z.; Li, H.; Jiang, W.; Wu, Z.; Liu, G.; Yao, Y.; et al. Performance and Experimental Progress of a Compact W-band High Average Power Sheet Beam Extended Interaction Oscillator. *IEEE Electron Device Lett.* **2023**, *44*, 144–147. [CrossRef]
16. Chang, Z.; Shu, G.; Tian, Y.; He, W. A Multimode Extended Interaction Oscillator with Broad Continuous Electric Tuning Range. *IEEE Trans. Electron Devices* **2022**, *69*, 3947–3952. [CrossRef]
17. Chang, Z.; Shu, G.; He, W. An Extended Interaction Oscillator Capable of Continuous Multimode Operation. *IEEE Trans. Electron Devices* **2021**, *68*, 6470–6475. [CrossRef]
18. Liao, J.; Shu, G.; Lin, G.; Lin, J.; Li, Q.; He, J.; Ren, J.; Chang, Z.; Xu, B.; Deng, J.; et al. Study of a 0.3-THz Extended Interaction Oscillator Based on the Pseudospark-Sourced Sheet Electron Beam. *IEEE Trans. Plasma Sci.* **2023**, *51*, 2199–2204. [CrossRef]
19. Bi, L.; Jiang, X.; Qin, Y.; Xu, C.; Wang, B.; Yin, Y.; Li, H.; Meng, L. Power Enhancement of Subterahertz Extended Interaction Oscillator Based on Overmoded Multigap Circuit and Linearly Distributed Two Electron Beams. *IEEE Trans. Electron Devices* **2022**, *69*, 792–797. [CrossRef]
20. He, X.; Yang, X.; Lu, G.; Yang, W.; Wu, F.; Yu, Z.; Jiang, J. Implementation of selective controlling electromagnetically induced transparency in terahertz graphene metamaterial. *Carbon* **2017**, *123*, 668–675. [CrossRef]

**Disclaimer/Publisher's Note:** The statements, opinions and data contained in all publications are solely those of the individual author(s) and contributor(s) and not of MDPI and/or the editor(s). MDPI and/or the editor(s) disclaim responsibility for any injury to people or property resulting from any ideas, methods, instructions or products referred to in the content.

Article

# Unveiling the Terahertz Nano-Fingerprint Spectrum of Single Artificial Metallic Resonator

Xingxing Xu [1,2], Fu Tang [1,2], Xiaoqiuyan Zhang [1,2,*] and Shenggang Liu [1,2]

[1] Terahertz Research Center, School of Electronic Science and Engineering, University of Electronic Science and Technology of China, Chengdu 611731, China; 202011022904@std.uestc.edu.cn (X.X.); 202112022427@std.uestc.edu.cn (F.T.); liusg@uestc.edu.cn (S.L.)
[2] Key Laboratory of Terahertz Technology, Ministry of Education, Chengdu 611731, China
* Correspondence: zhang_xqy@uestc.edu.cn

**Abstract:** As artificially engineered subwavelength periodic structures, terahertz (THz) metasurface devices exhibit an equivalent dielectric constant and dispersion relation akin to those of natural materials with specific THz absorption peaks, describable using the Lorentz model. Traditional verification methods typically involve testing structural arrays using reflected and transmitted optical paths. However, directly detecting the dielectric constant of individual units has remained a significant challenge. In this study, we employed a THz time-domain spectrometer-based scattering-type scanning near-field optical microscope (THz-TDS s-SNOM) to investigate the near-field nanoscale spectrum and resonant mode distribution of a single-metal double-gap split-ring resonator (DSRR) and rectangular antenna. The findings reveal that they exhibit a dispersion relation similar to that of natural materials in specific polarization directions, indicating that units of THz metasurface can be analogous to those of molecular structures in materials. This microscopic analysis of the dispersion relation of artificial structures offers new insights into the working mechanisms of THz metasurfaces.

**Keywords:** subwavelength; THz metasurface; THz-TDS s-SNOM; nanoscale spectrum

**Citation:** Xu, X.; Tang, F.; Zhang, X.; Liu, S. Unveiling the Terahertz Nano-Fingerprint Spectrum of Single Artificial Metallic Resonator. *Sensors* **2024**, *24*, 5866. https://doi.org/10.3390/s24185866

Academic Editor: Gintaras Valusis

Received: 1 August 2024
Revised: 30 August 2024
Accepted: 2 September 2024
Published: 10 September 2024

**Copyright:** © 2024 by the authors. Licensee MDPI, Basel, Switzerland. This article is an open access article distributed under the terms and conditions of the Creative Commons Attribution (CC BY) license (https://creativecommons.org/licenses/by/4.0/).

## 1. Introduction

THz metasurface devices have long been utilized in electromagnetic emission [1–3], control [4–7], imaging [8] and biochemical sensing [9–12], owing to their distinctive electromagnetic properties, with metallic resonant structures being among the most prevalent applications. The interaction between these structures and electromagnetic waves is predominantly explained using the LC circuit resonance model [13,14] and the equivalent dielectric constant model [15,16]. However, due to the sub-wavelength nature of metasurface elements, their interaction efficiency with THz waves is inherently low, necessitating the use of large-scale arrays to observe their influences on electromagnetic waves effectively. Through traditional THz far-field transmission and reflection experiments, researchers have demonstrated that the resonant absorption properties of arrays of artificial metal structures exhibit similarities to those of natural materials with specific absorption peaks in the THz frequency band, such as lactose [17], lead, and vermilion [18], the permittivity of which can be described by the Lorentz model. Nonetheless, artificial structures differ from natural materials in that their equivalent dielectric constant can be freely tuned by adjusting structural parameters, thereby offering broader application scenarios.

In THz metasurface devices, the electromagnetic characteristics of each unit can be significantly altered by micron-level parameter changes, thereby almost directly determining the electromagnetic properties of the entire array. The detection and analysis of individual elements are crucial for enhancing our understanding of the interaction mechanisms between metasurfaces and electromagnetic waves. Traditional far-field imaging methods are constrained by spatial resolution limitations, necessitating the diffraction limit to be broken to directly detect the equivalent dielectric constant or surface field distribution of a single metasurface element. The advent of THz near-field imaging technology has increased the imaging and spectral resolution of THz waves to the micron and even nanometer scale [19,20], enabling the direct measurement of nanoscale phenomena that have experienced rapid advancements in recent years, such as nanoresonators, surface plasmons [21], and phonon-polaritons [22]. Among these technologies, photoconductive probe, aperture, and scattering THz near-field imaging systems have reached considerable maturity in recent years. The fundamental principle of the first two methods involves positioning the detector at a distance of within a few or tens of microns from the sample surface for near-field detection [23,24]. Recent advancements have seen research teams utilizing these methods to achieve micron-level imaging and spectral formation of single-symmetric bimetallic antennas [25] and asymmetric resonant rings [26]. The scattering near-field system leverages an atomic force microscope (AFM) probe to scatter near-field information from the sample surface, thereby achieving imaging resolutions comparable to those of AFM [27–30]. Paul Dean's team has recently employed this system to observe the near-field surface resonance modes of single-rectangular-metallic antenna and ring structure [31]. Building on these studies, we designed a DSRR structure and a simple rectangular antenna, then employed THz-TDS s-SNOM to simultaneously obtain the THz broadband near-field intensity and phase information from its surface. This approach enabled the two-dimensional imaging of the surface resonance mode and the dispersion analysis of the equivalent dielectric constant in the resonance region. Additionally, by incrementally altering the position of one of the gaps, we observed notable differences between far-field and near-field detection.

## 2. Materials and Methods

### 2.1. Experimental Setup

The THz-TDS s-SNOM used in the experiment is a commercial product from Neaspec GmbH, and is composed of AFM and THz-TDS. The schematic diagram of the experimental optical path is illustrated in Figure 1. A collimated P-polarized THz pulse is obliquely incident and converged at the tip of the AFM probe, which operates in tapping mode, via an off-axis parabolic mirror with a focal length of 16 mm and an incident angle of approximately 52 degrees. In this configuration, the local field between the probe and the sample is modulated at the probe's tapping frequency and scattered into free space. Both the reflected and scattered signals are captured by a photoconductive antenna (PCA) receiver. When demodulating high-order near-field signals, which is carried out to extract relatively pure THz near-field information, the process is typically executed at each time delay point. In conventional far-field time-domain scanning, the signal at each time delay point remains relatively stable. However, in near-field systems, the signal at each time delay point encapsulates modulated scattered near-field information, rendering it inherently unstable. These weak near-field components can be extracted from the signal received by the PCA by utilizing a phase-locked demodulation system, which demodulates the higher-order harmonics of the probe's oscillation frequency. By scanning the delay line, the time-domain signal of the near-field can be incrementally reconstructed. The probe used in this setup is a 25PtIr500B-H50 type, customized by RMN Inc, the cantilever length of which exceeds 500 μm and the tip shank of which is about 80 μm, featuring a resonant frequency of 15.5 kHz and an amplitude of approximately 150 nm. To mitigate the influence of water vapor on the THz spectrum, the system operates in an environment with humidity maintained below 10%.

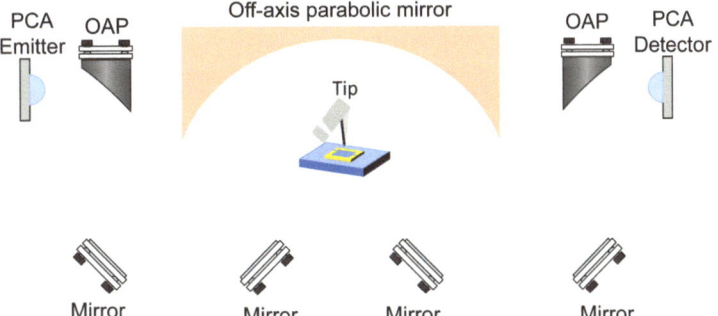

**Figure 1.** Schematic diagram of the THz-TDS s-SNOM.

## 2.2. Material Preparation

Figure 2a,b present scanning electron microscope (SEM) images of the sample and a schematic diagram of a single DSRR structure, respectively. The parameters of the designed DSRR are as follows: $p$ = 70 μm, L = 40 μm, w = 6 μm, and s = 2.5 μm. The horizontal distance between the center of the upper gap and the center of the structure, defined as the asymmetric coefficient, ac, is another critical parameter. In addition, 25 metallic rectangular antennas of different lengths were fabricated, each with a uniform width of 3 μm and lengths ranging from 38 μm to 86 μm in 2 μm increments. To minimize coupling effects and mutual interference between adjacent antennas, the inter-antenna spacing was designed to exceed 300 μm. In the experiment, these structures were fabricated from gold using a photolithography process, achieving a thickness of approximately 100 nm. The substrate material was high-resistance silicon (HR-Si) with a resistivity exceeding 10,000 Ω·cm and a thickness of 1 mm.

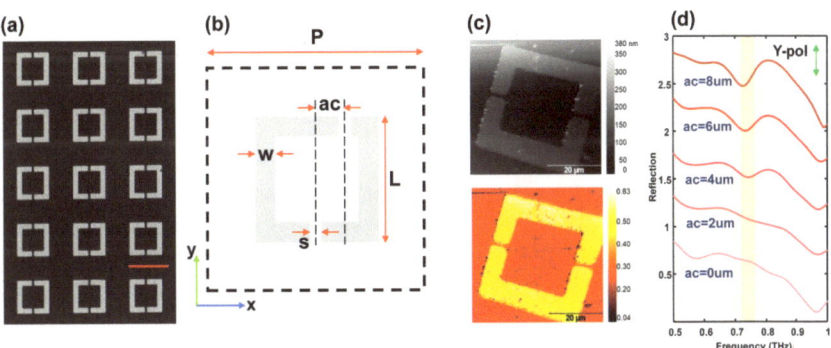

**Figure 2.** (**a**) SEM image of a symmetric DSRR array; the scale bar is 50 μm. (**b**) A schematic diagram of DSRR, in which ac is the asymmetric coefficient. (**c**) Topography (top) and 3rd-order near-field imaging of the time-domain peak (bottom). (**d**) The normalized far-field reflection spectrum of DSRR arrays with different asymmetric coefficients. We can see that the absorption peak increases and undergoes a redshift as the ac increases.

## 2.3. Methods of Detection

To achieve such high imaging resolution in the s-SNOM system, the most crucial aspect is the extraction of pure near-field signals amidst substantial background noise. Enhancing near-field scattering efficiency is typically accomplished by utilizing a metallic tip within the THz frequency band. The metallic tip significantly amplifies the local electric field at the probe tip due to its pronounced lightning rod effect, while simultaneously leveraging the mirror dipole interaction between the probe tip and the sample surface to scatter the surface electric field. Consequently, the spatial extent of the near-field electric field is primarily determined by the curvature radius of the probe tip. The quantization of the scattered signal can be effectively modeled via the polarization and radiation processes of the point dipole formed between the probe tip and the sample, commonly referred to as the point dipole model of the near field [32]. In this model, the probe tip is approximated as a sphere with radius $r$. As the probe oscillates with a certain frequency, $\Omega$, and amplitude, $A$, the minimum distance between the probe tip and the sample surface is $z$. Following the probe's oscillation, the distance, $d$, between the center of the probe tip and the sample surface can be expressed as follows:

$$d = z + r + A[1 + cos(2\pi\Omega t)]$$

At the same time, the polarizability, $\alpha$, of the probe tip can be expressed as follows:

$$\alpha = 4\pi r^3 (\varepsilon_t - 1)/(\varepsilon_t + 2)$$

In the aforementioned equation, $\varepsilon_t$ represents the dielectric constant of the probe material. When a P-polarized THz signal impinges upon the probe tip, the effective polarizability of the probe sample dipole is given by the following:

$$\alpha_{eff} = \alpha / \left(1 - \alpha\beta / \left(16\pi d^3\right)\right)$$

where $\beta$ is related to the sample reflection coefficient, which is calculated via $\beta = (\varepsilon_s - 1)/(\varepsilon_s + 1)$, including $\varepsilon_s$ for the dielectric constant of the sample. Combined with the reflection coefficient, $\gamma$, of the sample surface and the incident electric field intensity, $E_i$, the final scattered signal, $E_s$, of the probe can be expressed as follows:

$$E_s \propto \alpha_{eff}(1 + \gamma)^2 E_i$$

The above equation represents the modulation process of the probe on the near-field signal, where demodulation involves determining the Fourier expansion coefficients of the effective polarizability. As the probe retracts from the sample surface, the local field between the probe and the sample undergoes significant variation, decreasing exponentially with increasing distance from the sample. In contrast, the reflected background signal decreases almost linearly with increasing distance due to the large spot size. Therefore, theoretically, higher-order demodulation yields purer near-field signal extraction. However, due to the characteristics of Fourier expansion in exponentially decaying signals, higher-order near-field signals are inherently weaker, resulting in a lower signal-to-noise ratio (SNR) in practical experiments. Thus, it is crucial to select an appropriate order based on the specific experimental conditions.

### 2.4. Data Processing

Due to the normalization of the THz pulse being influenced by the time delay and the unknown flatness of the sample surface, a substrate region within tens of micrometers near the structure is typically selected when acquiring the reference near-field signal. This approach ensures a minimal delay difference between the sample and the reference signal. During the time-domain near-field signal scan, each delay point is integrated for 200 ms to achieve a sufficient SNR for the 2nd-order signal. While the 3rd-order near-field signal can also be observed under these conditions, the reference signal generally exhibits a relatively lower SNR. The normalization uses the following method:

$$\text{Norm } S_2(\omega) = \left| \frac{S_{2,sample}(\omega)}{S_{2,ref}(\omega)} \right|$$

$$\text{Norm } P_2(\omega) = P_{2,sample}(\omega) - P_{2,ref}(\omega) \quad (1)$$

where $S_{2,sample}$ and $P_{2,sample}$ are the 2nd-order near-field amplitude and phase obtained in the surface of our DSRR, respectively. Also, $S_{2,ref}$ and $P_{2,ref}$ are the same kinds of data obtained in the surface of substrate, which is HR-Si. In phase processing, it may be necessary to subtract the baseline. In addition, the far-field reflection spectrum of periodic structures in the following figures should be normalized to the reflection spectrum of the pure substrate, which is also HR-Si.

## 3. Results

Initially, AFM topography and third-order near-field time-domain peak imaging were performed on a single symmetric DSRR with an asymmetric coefficient of ac = 0 μm, which landed in the center area of the array. The results are presented in Figure 2c. Due to the scanning range approaching the maximum imaging region of the AFM (57 μm × 66 μm), some distortions are observed at the edges of the topograph and near-field images. Nevertheless, these distortions do not impact the analysis of the near-field signal strength. It is evident that the near-field signal from gold is significantly higher than that from high-resistance silicon, while the near-field signal strength from gold remains nearly uniform. To ascertain the distribution of the resonance mode, it is imperative to collect the time-domain signal at each point and image the single-frequency point. To determine the actual resonant frequency of the designed DSRR, the far-field reflection spectrum of the periodic structure was measured. For comparative purposes, array structures with varying asymmetric coefficients (ac = 0, 2, 4, 6, and 8 μm) were fabricated on the same silicon substrate. Each array comprised 100 elements (10 × 10) spanning an area of 700 μm × 700 μm, with an inter-array spacing exceeding 1.3 mm (see Figure A1), ensuring that the THz spot illuminated only one array at a time. In the experiment, the far-field reflection and near-field tests share the same optical path, allowing the far-field measurements to reliably predict the near-field resonance frequency. The far-field reflection spectra for different asymmetric coefficients are depicted in Figure 2d. These results indicate that all absorption peaks are centered around 0.73 THz ($\lambda \approx 6p$), with larger asymmetric coefficients corresponding to lower absorption peak frequencies. This observation suggests that lower frequencies correlate with the resonance of a longer arm, implying that the absorption peak is predominantly influenced by the long arm of the structure.

According to the antenna effect of the THz probe [33,34] and the finite dipole model theory [32], the probe in the THz near-field system primarily couples and scatters the external component of the sample surface. The $E_z$ contour plot and surface current distribution of the asymmetric DSRR at the resonant frequency, obtained through CST simulation, are presented in Figure 3b. We can see that the surface current forms a closed loop, indicating that the resonant mode corresponds to a magnetic dipole mode. Additionally, the electric field resonance is predominantly concentrated near the vicinity of the split gap, highlighting the localization of the electromagnetic field in that region. Figure 3a illustrates a schematic diagram of the optical path used in our actual test. The THz wave is obliquely incident from the left side, with the probe positioned on a metal arm near the opening, specifically at the red dot indicated in Figure 3b. For an asymmetric coefficient of ac = 0 μm, the second-order time-domain near-field signal and corresponding normalized second-order near-field intensity and phase spectra with a time window of 2–10 ps measured at the red dot are shown in Figure 3c. Approximately 4 ps behind the main peak (indicated by the gray dashed line box), a peak misalignment between the substrate and the DSRR begins to manifest, leading to the subsequent spectral differences. We can see that the intensity and phase spectra align closely with the real and imaginary parts of the dielectric constant, as described by the typical Lorentz model. In 2019, Haewook Han's team reported near-field observations of α-lactose and β-lactose, calculating the equivalent permittivity of these lactose molecules using the probe sample line dipole model and HDPE as references, achieving a good fit with the Lorentz model [35]. Similar structures have been tested in the infrared near-field for organic materials such as polystyrene (PS) and polymethyl-methacrylate (PMMA) [36]. The near-field model reveals that the strength of the near-field signal is positively correlated with the real part of the dielectric constant, and the phase of the near-field signal is positively correlated with the imaginary part of the dielectric constant. Consequently, the equivalent dielectric constant extracted from the near-field information at the electric field resonance locations on the metasurface is consistent with the Lorentz model. Using the Lorentz model formula,

$$\varepsilon(\omega) = \varepsilon_\infty + \frac{\omega_p^2}{\omega_0^2 - \omega^2 - i\gamma\omega} \qquad (2)$$

a preliminary fit can be applied to the intensity and phase of the second-order near-field normalization using the Lorentz model. In the equation, $\omega_0$ and $\gamma$ represent the resonant frequency and damping parameters, respectively. The fitting results are depicted by the dashed line in Figure 3c, with the resonant frequency, according to the fitting coefficients, these also being 0.73 THz. To further elucidate the resonant size-sensitive properties of the metasurface elements, we varied the asymmetric coefficients of the DSRR and conducted near-field experiments on individual structures. The second-order normalized near-field phase spectra for different asymmetric coefficients are shown in Figure 3d. The testing positions remained consistent, revealing significant changes in the resonant frequency of a single resonant ring. Although there are minor discrepancies between the resonant frequencies and the positions of the far-field absorption peaks, the trend is consistent with that observed in the far-field measurements.

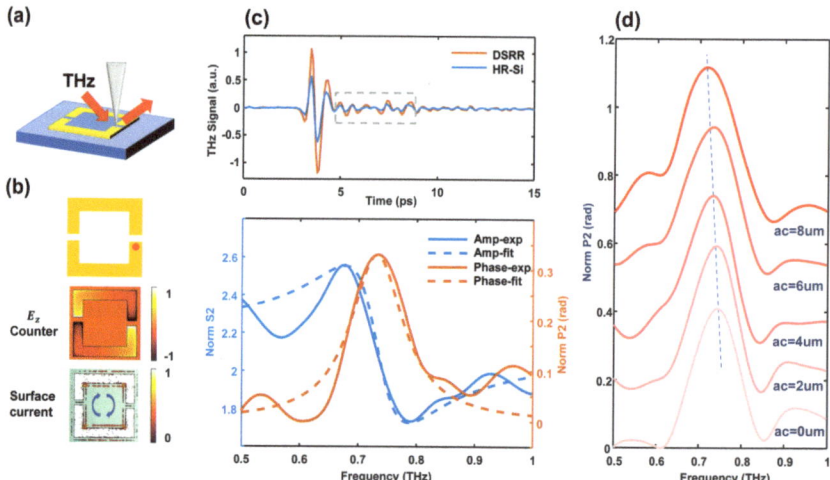

**Figure 3.** (**a**) A schematic diagram of our near-field experiment. (**b**) A simple diagram of the symmetric DSRR; the red dot indicates where the tip scatters the near-field signal (top). The contour of the z component of the electric field (center) and surface current (bottom) as ac = 2 μm. (**c**) The top picture shows the 2nd-order near-field time-domain signal measured at the base and at the red dot, and the bottom shows the 2nd-order near-field spectrum of the DSRR normalized to the HR-Si substrate, where the real and dashed lines correspond to the experimental and Lorentz fitting results, respectively. (**d**) The normalized 2nd-order near-field phase spectrum measured at the red dot with the increasing ac from the bottom up.

As a typical rectangular antenna within a metallic metasurface structure, although the resonance phenomena are less pronounced compared to those of resonant rings, it still enables efficient tuning of the resonant frequency. Additionally, an isolated metallic rectangular antenna typically functions as a half-wave antenna, with its resonant frequency primarily dependent on its physical length and the dielectric constant of the substrate material. The resonant wavelength exhibits direct proportionality to the antenna's length, facilitating the analysis of the experimental results. The optical image of some metallic rectangular antennas is shown in Figure 4a, and Figure 4b illustrates the antenna's structural diagram (top) and the corresponding simulation contour diagram at the resonant frequency (bottom), where the electric field in the simulation is an out-of-plane component, consistent with the characteristic resonant mode distribution of a half-wave antenna. Figure 4e presents the optical path diagram for both the simulation and the actual experiment. To conduct a more detailed analysis of the antenna's resonance, we initially compared the normalized second-order near-field time-domain signals of metallic antennas with varying lengths, as shown in Figure 4c. Given that the water vapor content in the experimental environment was below 3%, the near-field time-domain signals in the pure substrate exhibited only the primary and secondary reflections resulting from surface wave propagation on the probe [37]. When the near-field signal was collected at the edge of the metal antenna (indicated by the red dot in Figure 4b), a clear observation could be made of the electric field oscillation occurring subsequent to the main peak in the near-field signal. This oscillation persisted until the reflected signal from the probe cantilever's end reached the probe tip and was scattered, corresponding to the reflection time in the time domain. For longer rectangular antennas, it was observed that the period of electromagnetic oscillation increased correspondingly, directly confirming the resonant behavior of THz waves on metallic antennas through time-domain analysis.

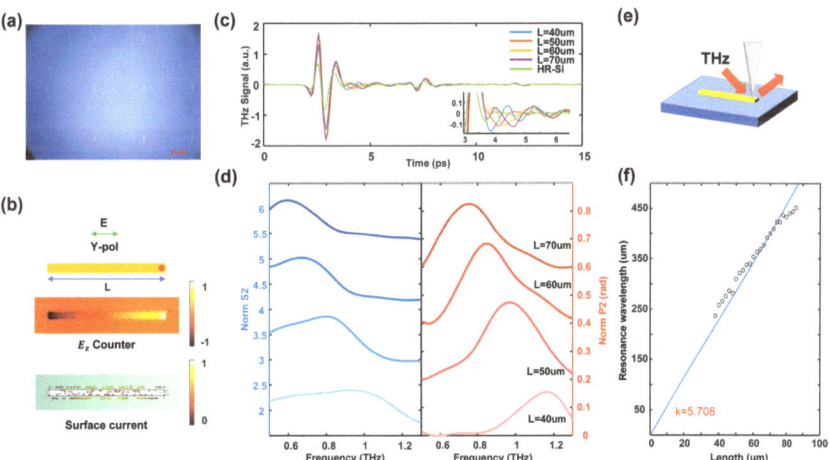

**Figure 4.** (**a**) An optical image of metallic rectangular antennas; the scale bar is 50 μm. (**b**) A simple diagram of a rectangular antenna; the red dot indicates where the tip scatters the near-field signal (top). A contour picture of the z component of the electric field (center) and surface current (bottom) as L = 60 μm. (**c**) The 2nd-order near-field time domain signal measured at the base and at the red dot when L = 40, 50, 60 and 70 μm. The embedded picture is an enlarged graph of the signal. (**d**) The corresponding 2nd-order near-field amplitude and phase spectrum normalized to those of the HR-Si substrate. (**e**) A schematic diagram of our near-field experiment. (**f**) The experimentally measured resonant wavelengths of rectangular antennas with different lengths (black circles) and corresponding linear fitting results (blue solid lines).

Based on the time-domain signal, it is possible to predict the corresponding characteristics in the frequency domain. Figure 4d presents the FFT results of the time-domain signal shown in Figure 4c. In this analysis, a signal within the time window from 1 ps to 6.5 ps—spanning from the main peak to the primary reflection—was selected. However, due to the prolonged duration of resonance in the metal antenna, this method may have resulted in a reduced measured near-field resonance Q-factor. The choice of different time windows for signal extraction introduced minor variations in the measured resonance frequency (refer to Figure A2b), but it did not affect the overall linearity of the resonant spectrum or the trend of resonant frequency changes for metal antennas of varying lengths. In addition, the resonant spectra could still be accurately modeled using the Lorentzian profile. We observed that the trend of resonant frequency corresponding to metal antennas of different lengths aligned with our predictions. Additionally, the relationship between the resonant wavelength and the lengths of all 25 antennas is plotted as a scatter plot in Figure 4f, with a linear fit applied. The slope of the fitted line was determined to be 5.708, allowing us to approximate that the THz propagation velocity in a metal with a high-resistance silicon substrate, whose reflective index is 3.4, is approximately $c/2.854$. The dielectric properties of the engineered metasurface elements were further validated through near-field spectral testing of the metallic rectangular antenna.

To obtain the resonant field distribution on the surface of the DSRR, time-domain scanning was performed along the metal edge of the DSRR with a scanning interval of 1 μm. By performing FFT for each point and normalizing it to the substrate, a phase contour diagram for a single frequency point, as shown in Figure 5c, was obtained (the phase of substrate is defined as 0 rad). To eliminate inter-structure interactions within the array, this DSRR was positioned in an isolated location with no other structures within a 400 μm radius. The corresponding frequency gradually increased from bottom to top, with 0.73 THz being the resonant frequency of the DSRR. It can be observed that the electric field of the resonance is primarily concentrated on both sides of the opening and gradually weakens as it moves away from the opening. The resonance disappears at the 2.5 μm opening, fully demonstrating the spatial resolution of the near-field scattering. Figure 5a,b show the near-field phase spectra obtained via scanning along the red and green dashed lines of the DSRR in Figure 5c, respectively. It can be intuitively observed that there is essentially no resonant electric field in the central region of the opening and the unopened edges. For the metallic rectangular antenna, we also conducted near-field spectrum tests at different locations, the results of which are basically consistent with the simulation results. The details are shown in Figure A2a.

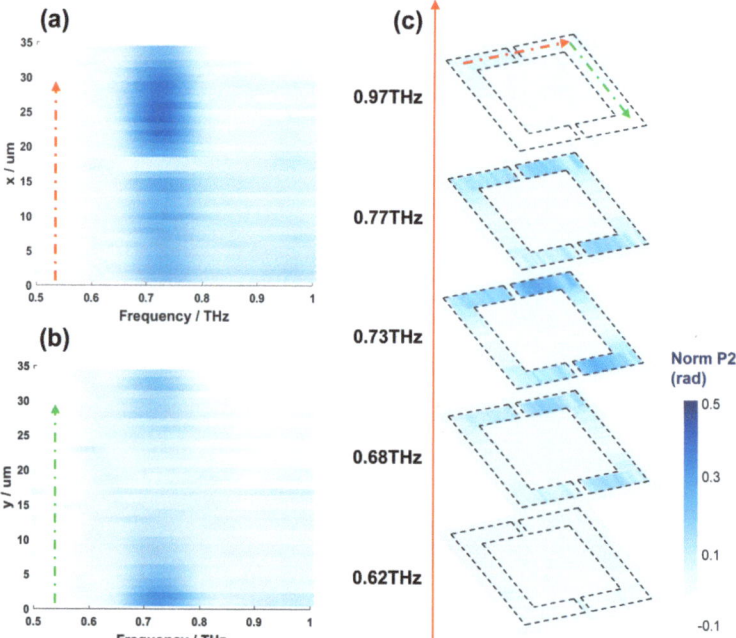

**Figure 5.** (a) The line scanning the normalized 2nd-order near-field phase spectrum of the opening side, where the scanning direction and range correspond to the red dashed line in (c). (b) The same as (a), except the scanning direction and range correspond to the green dashed line in (c). (c) Normalized 2nd-order near-field phase contour diagram with different frequencies (the resonant frequency of this DSRR is 0.73 THz).

## 4. Discussion

Although we observed the field distribution and nanoscale spectrum on the surface of the metasurface element, when we pay attention to the simulation results (see Figure 3b) and the experimental results (see Figure 5c), it can be found that a phase difference between the two sides of the opening was not detected, and the same result can also be observed in metallic rectangular antennas (see Figures 4b and A2a), which may be caused by the disturbance of the surface electric field by the probe. The metal probe's lightning rod effect causes a large accumulation of charge at the tip [38], forming a strong local electric field. Consequently, when the probe is near the surface of the structure, some coupling between the local field and the resonant field on the surface of the structure inevitably occurs, affecting the intensity and phase information of the scattered field. This phenomenon was also noted by other research teams in previous studies [39].

Given that the resonant mode of the DSRR corresponds to a magnetic dipole, the entire ring effectively constitutes a resonant wavelength. Consequently, the measured results should approximate the resonant frequency of an 80 μm antenna in a half-wave antenna configuration. As observed in Figure 3, the resonant frequency of the DSRR is approximately 0.72 THz, which closely aligns with the resonant frequency of the 76 μm rectangular antenna. Considering the inherent uncertainties in structural fabrication and experimental measurement, it is reasonable to conclude that the two structures exhibit a strong correspondence in their resonant behaviors.

Additionally, it is noteworthy that as we reduce the asymmetry coefficient, the absorption peak in the far-field reflection spectrum continuously diminishes (see Figure 2d). In the case of structural symmetry, the absorption peak is almost imperceptible. Even when the sample is rotated by 90 degrees, no significant far-field absorption peak is observed (as shown in Figure 6). However, near-field tests reveal a strong resonance near 0.73 THz, indicative of a typical radiation interference cancellation phenomenon. Previous studies by Longqing Cong's team demonstrated that symmetric structures produce the bound states in the continuum (BIC) phenomenon under a normally incident Y-polarized electric field due to the radiation interference of two symmetric metal arms [40]. Although our experiment involves oblique incidence, the two metal arms remain symmetric relative to the incident THz wave, causing interference cancellation. When the asymmetry coefficient increases and the structure loses its symmetry, the interference cancellation condition is no longer met, and the structure's resonance is reflected in the far-field absorption peak.

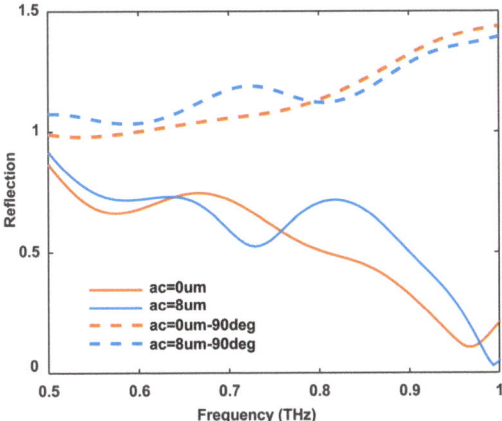

**Figure 6.** Normalized far-field reflection spectrum of symmetric DSRR array at two angles perpendicular to each other. The solid lines are the same as those in Figure 2d, and the dashed lines are measured by rotating the sample by 90 degrees. When ac is 0 μm, the array has no apparent absorption in both conditions. Since the reference signal comes from the HR-Si substrate, it is reasonable for the normalized reflection of the periodic structures to be greater than 1.

## 5. Conclusions

In conclusion, we employed THz-TDS s-SNOM to conduct nanoscale spectrum testing of a single metallic DSRR and to achieve sub-micron imaging of its resonant mode. By extracting the near-field spectrum and normalizing it to that of high-resistance silicon substrate, we found that the intensity and phase conformed to the typical Lorentz model, aligning closely with results obtained for natural resonant materials. At the same time, the near-field spectrum of rectangular metal antennas of different lengths was observed, and the propagation speed of THz in metal with silicon substrate was estimated roughly. Additionally, we simulated both resonant and non-resonant frequencies on the surface of the structure with a sub-micron resolution, directly observing that the resonant field was concentrated on both sides of the opening and decayed towards the metal center. For the symmetric DSRR, the persistence of near-field resonance indicates that the disappearance of the absorption peak in the far-field reflection is due to interference cancellation of structural radiation. These testing methodologies can be broadly applied to various metallic metasurface structures. Our findings offer novel insights into the working mechanisms of metasurface units at a microscopic level and are anticipated to advance the development of THz emission, regulation, and sensing technologies.

**Author Contributions:** Conceptualization, X.X.; methodology, X.X. and F.T.; formal analysis, X.X. and X.Z.; investigation, X.X. and F.T.; data curation, X.X. and F.T.; writing—original draft preparation, X.X.; writing—review and editing, X.X., S.L. and X.Z. All authors have read and agreed to the published version of the manuscript.

**Funding:** This work is supported by the National Key Research and Development Program of China under Grant 2020YFA0714001, the Natural Science Foundation of China under Grant 61988102, 61921002 and 62071108, the Fundamental Research Funds for the Central Universities under Grant ZYGX2020ZB007, the Sichuan Science and Technology Program (2023NSFSC1477), and the fund of Key Laboratory of THz Technology, Ministry of Education, China.

**Institutional Review Board Statement:** Not applicable.

**Informed Consent Statement:** Not applicable.

**Data Availability Statement:** The data presented in this study are available at the links mentioned in the text or on request from the corresponding author.

**Conflicts of Interest:** The authors declare no conflicts of interest.

## Appendix A

These pictures present the optical micrographs of the fabricated object, the near-field spectrum at various positions along the rectangular antenna, and a comparison of near-field spectra derived from different time-domain intervals. The measurement results for the rectangular antenna and DSRR exhibit a commonality: the anticipated phase-opposite behavior, as predicted via simulations, was not observed at both edges. For an analysis of this discrepancy, please refer to the Section 4. Additionally, it was noted that reflections originating from the probe's tip had a minimal impact on the near-field measurement of the resonant frequency.

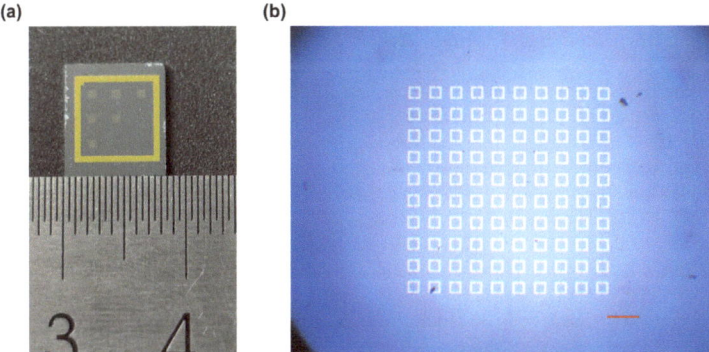

**Figure A1.** (**a**) Physical diagram of processed DSRR sample. Six DSRR arrays with six asymmetric coefficients are processed in the same high-resistance silicon wafer, and the region with the single DSRR is in the lower right corner. (**b**) Optical images of the rDSRR array at ac = 0 µm; the scale bar is 100 µm.

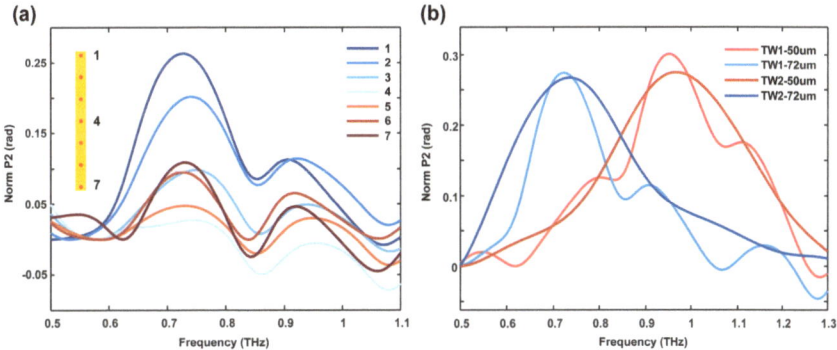

**Figure A2.** (**a**) Normalized 2nd-order near-field phase spectrum of 7 points uniformly acquired on a metallic rectangular antenna with L = 72 µm. The corresponding acquisition position of each spectrum is shown in the embedded figure. (**b**) Normalized 2nd-order near-field phase spectrum comparison under two time window interceptions, where time window 1 is from 1 to 10 ps, and time window 2 is from 1 to 6.5 ps.

# References

1. Liu, Y.; Bai, Z.; Xu, Y.; Wu, X.; Sun, Y.; Li, H.; Sun, T.; Kong, R.; Pandey, C.; Kraft, M.; et al. Generation of tailored terahertz waves from monolithic integrated metamaterials onto spintronic terahertz emitters. *Nanotechnology* **2020**, *32*, 105201. [CrossRef] [PubMed]
2. Yu, X.; Suzuki, Y.; Van Ta, M.; Suzuki, S.; Asada, M. Highly Efficient Resonant Tunneling Diode Terahertz Oscillator With a Split Ring Resonator. *IEEE Electron Device Lett.* **2021**, *42*, 982–985. [CrossRef]
3. Pu, M.; Luo, X. Advancing nonlinear nanophotonics: Harnessing membrane metasurfaces for third-harmonic generation and imaging. *Opto-Electron. Adv.* **2023**, *6*, 230153. [CrossRef]
4. Degl'Innocenti, R.; Lin, H.; Navarro-Cia, M. Recent progress in terahertz metamaterial modulators. *Nanophotonics* **2022**, *11*, 1485–1514. [CrossRef]
5. Li, J.; Wang, G.; Yue, Z.; Liu, J.; Li, J.; Zheng, C.; Zhang, Y.; Zhang, Y.; Yao, J. Dynamic phase assembled terahertz metalens for reversible conversion between linear polarization and arbitrary circular polarization. *Opto-Electron. Adv.* **2022**, *5*, 210062. [CrossRef]
6. Zhao, H.; Wang, X.; Liu, S.; Zhang, Y. Highly efficient vectorial field manipulation using a transmitted tri-layer metasurface in the terahertz band. *Opto-Electron. Adv.* **2023**, *6*, 220012. [CrossRef]
7. Gao, H.; Fan, X.; Liu, S.;Wang, Y.; Liu, Y.; Wang, X.; Xu, K.; Deng, L.; Zeng, C.; Li, T.; et al. Multi-foci metalens for spectra and polarization ellipticity recognition and reconstruction. *Opto-Electron. Sci.* **2023**, *2*, 220026. [CrossRef]
8. Li, Y.; Huang, X.; Liu, S.; Liang, H.; Ling, Y.; Su, Y. Metasurfaces for near-eye display applications. *Opto-Electron. Sci.* **2023**, *2*, 230025. [CrossRef]

9. Singh, R.; Cao, W.; Al-Naib, I.; Cong, L.; Withayachumnankul, W.; Zhang, W. Ultrasensitive terahertz sensing with high-Q Fano resonances in metasurfaces. *Appl. Phys. Lett.* **2014**, *105*, 171101. [CrossRef]
10. Srivastava, Y.K.; Ako, R.T.; Gupta, M.; Bhaskaran, M.; Sriram, S.; Singh, R. Terahertz sensing of 7 nm dielectric film with bound states in the continuum metasurfaces. *Appl. Phys. Lett.* **2019**, *115*, 151105. [CrossRef]
11. Wang, R.; Xu, L.; Huang, L.; Zhang, X.; Ruan, H.; Yang, X.; Lou, J.; Chang, C.; Du, X. Ultrasensitive Terahertz Biodetection Enabled by Quasi-BIC-Based Metasensors. *Small* **2023**, *19*, e2301165. [CrossRef] [PubMed]
12. Cui, F.; Huang, X.; Zhou, Q.; Tong, Y.; Liu, F.; Tang, Y.; Meng, X.; Li, C.; Fang, B.; Jing, X. Magnetic toroidal dipole resonance terahertz wave biosensor based on all-silicon metasurface. *Opt. Lasers Eng.* **2024**, *177*, 108128. [CrossRef]
13. Xiong, H.; Li, X.-M. Parametric Investigation and Analysis of an Electric-LC Resonator by Using LC Circuit Model. *Appl. Comput. Electromagn. Soc.* **2020**, *35*, 1113–1118. [CrossRef]
14. Chen, H.S.; Ran, L.X.; Huangfu, J.T.; Grzegorczyk, T.M.; Kong, J.A. Equivalent circuit model for left-handed metamaterials. *J. Appl. Phys.* **2006**, *100*, 024915. [CrossRef]
15. Smith, D.R.; Vier, D.C.; Koschny, T.; Soukoulis, C.M. Electromagnetic parameter retrieval from inhomogeneous metamaterials. *Phys. Rev. E* **2005**, *71*, 036617. [CrossRef]
16. Zhou, J.F.; Zhang, L.; Tuttle, G.; Koschny, T.; Soukoulis, C.M. Negative index materials using simple short wire pairs. *Phys. Rev. B* **2006**, *73*, 041101. [CrossRef]
17. Datta, S.; Prasertsuk, K.; Khammata, N.; Rattanawan, P.; Chia, J.Y.; Jintamethasawat, R.; Chulapakorn, T.; Limpanuparb, T. Terahertz Spectroscopic Analysis of Lactose in Infant Formula: Implications for Detection and Quantification. *Molecules* **2022**, *27*, 5040. [CrossRef] [PubMed]
18. Lee, J.E.; Lee, H.; Kim, J.; Jung, T.S.; Kim, J.H.; Kim, J.; Baek, N.Y.; Song, Y.N.; Lee, H.H.; Kim, J.H. Terahertz Spectroscopic Analysis of the Vermilion Pigment in Free-Standing and Polyethylene-Mixed Forms. *ACS Omega* **2021**, *6*, 13802–13806. [CrossRef]
19. Cocker, T.L.; Jelic, V.; Hillenbrand, R.; Hegmann, F.A. Nanoscale terahertz scanning probe microscopy. *Nat. Photonics* **2021**, *15*, 558–569. [CrossRef]
20. Chen, X.; Hu, D.; Mescall, R.; You, G.; Basov, D.N.; Dai, Q.; Liu, M. Modern Scattering-Type Scanning Near-Field Optical Microscopy for Advanced Material Research. *Adv. Mater.* **2019**, *31*, e1804774. [CrossRef]
21. Chen, S.; Leng, P.L.; Konecna, A.; Modin, E.; Gutierrez-Amigo, M.; Vicentini, E.; Martin-Garcia, B.; Barra-Burillo, M.; Niehues, I.; Escudero, C.M.; et al. Real-space observation of ultraconfined in-plane anisotropic acoustic terahertz plasmon polaritons. *Nat. Mater.* **2023**, *22*, 860–866. [CrossRef] [PubMed]
22. Yuan, Z.; Chen, R.K.; Li, P.N.; Nikitin, A.Y.; Hillenbrand, R.; Zhang, X.L. Extremely Confined Acoustic Phonon Polaritons in Monolayer-hBN/Metal Heterostructures for Strong Light-Matter Interactions. *ACS Photonics* **2020**, *7*, 2610–2617. [CrossRef]
23. van Hoof, N.J.J.; Abujetas, D.R.; Ter Huurne, S.E.T.; Verdelli, F.; Timmermans, G.C.A.; Sanchez-Gil, J.A.; Rivas, J.G. Unveiling the Symmetry Protection of Bound States in the Continuum with Terahertz Near-Field Imaging. *ACS Photonics* **2021**, *8*, 3010–3016. [CrossRef]
24. ter Huurne, S.; Abujetas, D.R.; van Hoof, N.; Sanchez-Gil, J.A.; Gómez Rivas, J. Direct Observation of Lateral Field Confinement in Symmetry-Protected THz Bound States in the Continuum. *Adv. Opt. Mater.* **2023**, *11*, 2202403. [CrossRef]
25. Hale, L.L.; Keller, J.; Siday, T.; Hermans, R.I.; Haase, J.; Reno, J.L.; Brener, I.; Scalari, G.; Faist, J.; Mitrofanov, O. Noninvasive Near-Field Spectroscopy of Single Subwavelength Complementary Resonators. *Laser Photonics Rev.* **2020**, *14*, 1900254. [CrossRef]
26. Lu, Y.; Hale, L.L.; Zaman, A.M.; Addamane, S.J.; Brener, I.; Mitrofanov, O.; Degl'Innocenti, R. Near-Field Spectroscopy of Individual Asymmetric Split-Ring Terahertz Resonators. *ACS Photonics* **2023**, *10*, 2832–2838. [CrossRef]
27. Maissen, C.; Chen, S.; Nikulina, E.; Govyadinov, A.; Hillenbrand, R. Probes for Ultrasensitive THz Nanoscopy. *ACS Photonics* **2019**, *6*, 1279–1288. [CrossRef]
28. Zeng, Y.; Lu, D.; Xu, X.; Zhang, X.; Wan, H.; Wang, J.; Jiang, X.; Yang, X.; Xu, M.; Wen, Q.; et al. Laser-Printed Terahertz Plasmonic Phase-Change Metasurfaces. *Adv. Opt. Mater.* **2023**, *11*, 2202651. [CrossRef]
29. Degl'Innocenti, R.; Wallis, R.; Wei, B.; Xiao, L.; Kindness, S.J.; Mitrofanov, O.; Braeuninger-Weimer, P.; Hofmann, S.; Beere, H.E.; Ritchie, D.A. Terahertz Nanoscopy of Plasmonic Resonances with a Quantum Cascade Laser. *ACS Photonics* **2017**, *4*, 2150–2157. [CrossRef]
30. Thomas, L.; Hannotte, T.; Santos, C.N.; Walter, B.; Lavancier, M.; Eliet, S.; Faucher, M.; Lampin, J.F.; Peretti, R. Imaging of THz Photonic Modes by Scattering Scanning Near-Field Optical Microscopy. *ACS Appl. Mater. Interfaces* **2022**, *14*, 32608–32617. [CrossRef]
31. Sulollari, N.; Keeley, J.; Park, S.; Rubino, P.; Burnett, A.D.; Li, L.; Rosamond, M.C.; Linfield, E.H.; Davies, A.G.; Cunningham, J.E.; et al. Coherent terahertz microscopy of modal field distributions in micro-resonators. *APL Photonics* **2021**, *6*, 066104. [CrossRef]
32. Cvitkovic, A.; Ocelic, N.; Hillenbr, R. Analytical model for quantitative prediction of material contrasts in scattering-type near-field optical microscopy. *Opt. Express* **2007**, *15*, 8550–8565. [CrossRef] [PubMed]
33. Mastel, S.; Lundeberg, M.B.; Alonso-González, P.; Gao, Y.; Watanabe, K.; Taniguchi, T.; Hone, J.; Koppens, F.H.; Nikitin, A.Y.; Hillenbr, R. Terahertz Nanofocusing with Cantilevered Terahertz-Resonant Antenna Tips. *Nano Lett.* **2017**, *17*, 6526–6533. [CrossRef] [PubMed]
34. Siday, T.; Natrella, M.; Wu, J.; Liu, H.; Mitrofanov, O. Resonant terahertz probes for near-field scattering microscopy. *Opt. Express* **2017**, *25*, 27874–27885. [CrossRef]

35. Moon, K.; Do, Y.; Park, H.; Kim, J.; Kang, H.; Lee, G.; Lim, J.H.; Kim, J.W.; Han, H. Computed terahertz near-field mapping of molecular resonances of lactose stereo-isomer impurities with sub-attomole sensitivity. *Sci. Rep.* **2019**, *9*, 16915. [CrossRef]
36. Mester, L.; Govyadinov, A.A.; Chen, S.; Goikoetxea, M.; Hillenbr, R. Subsurface chemical nanoidentification by nano-FTIR spectroscopy. *Nat. Commun.* **2020**, *11*, 3359. [CrossRef]
37. Hu, M.; Zhang, X.; Zhang, X.; Zhang, Z.; Zhang, T.; Xu, X.; Tang, F.; Yang, J.; Wang, J.; Jiang, H.; et al. Time-domain-filtered terahertz nanoscopy of intrinsic light-matter interactions. *Res. Sq.* **2023**. . [CrossRef]
38. McLeod, A.S.; Kelly, P.; Goldflam, M.D.; Gainsforth, Z.; Westphal, A.J.; Dominguez, G.; Thiemens, M.H.; Fogler, M.M.; Basov, D.N. Model for quantitative tip-enhanced spectroscopy and the extraction of nanoscale-resolved optical constants. *Phys. Rev. B* **2014**, *90*, 085136. [CrossRef]
39. Caselli, N.; Intonti, F.; La China, F.; Biccari, F.; Riboli, F.; Gerardino, A.; Li, L.; Linfield, E.H.; Pagliano, F.; Fiore, A.; et al. Generalized Fano lineshapes reveal exceptional points in photonic molecules. *Nat. Commun.* **2018**, *9*, 396. [CrossRef]
40. Cong, L.; Singh, R. Symmetry-Protected Dual Bound States in the Continuum in Metamaterials. *Adv. Opt. Mater.* **2019**, *7*, 1900383. [CrossRef]

**Disclaimer/Publisher's Note:** The statements, opinions and data contained in all publications are solely those of the individual author(s) and contributor(s) and not of MDPI and/or the editor(s). MDPI and/or the editor(s) disclaim responsibility for any injury to people or property resulting from any ideas, methods, instructions or products referred to in the content.

Article

# A Symmetrical Quasi-Synchronous Step-Transition Folded Waveguide Slow Wave Structure for 650 GHz Traveling Wave Tubes

Duo Xu [1], Tenglong He [2], Yuan Zheng [1], Zhigang Lu [1], Huarong Gong [1], Zhanliang Wang [1], Zhaoyun Duan [1] and Shaomeng Wang [1,*]

[1] School of Electronic Science and Engineering, University of Electronic Science and Technology of China, No. 2006, Xiyuan Ave., West Hi-Tech Zone, Chengdu 611731, China; d.xu@uestc.edu.cn (D.X.); zyzheng@uestc.edu.cn (Y.Z.); lzhgchnn@uestc.edu.cn (Z.L.); hrgong@uestc.edu.cn (H.G.); wangzl@uestc.edu.cn (Z.W.); zhyduan@uestc.edu.cn (Z.D.)
[2] Southwest Electronic Equipment Research Institute, No. 496, Yingkang West Road, Jinniu Distinct, Chengdu 610036, China; faithhill@foxmail.com
* Correspondence: wangsm@uestc.edu.cn

**Abstract:** For the purpose of improving performance and reducing the fabrication difficulty of terahertz traveling wave tubes (TWTs), this paper proposes a novel single-section high-gain slow wave structure (SWS), which is named the symmetrical quasi-synchronous step-transition (SQSST) folded waveguide (FW). The SQSST-FW SWS has an artificially designed quasi-synchronous region (QSR) to suppress self-oscillations for sustaining a high gain in an untruncated circuit. Simultaneously, a symmetrical design can improve the efficiency performance to some extent. A prototype of the SQSST-FW SWS for 650 GHz TWTs is designed based on small-signal analysis and numerical simulation. The simulation results indicate that the maximum saturation gain of the designed 650 GHz SQSST-FW TWT is 39.1 dB in a 34.3 mm slow wave circuit, occurring at the 645 GHz point when a 25.4 kV 15 mA electron beam and a 0.43 mW sinusoidal input signal are applied. In addition, a maximum output power exceeding 4 W is observed at the 648 GHz point using the same beam with an increased input power of around 2.8 mW.

**Keywords:** self-oscillation suppressing; terahertz radiation; traveling wave tubes

## 1. Introduction

Being at the overlap of electronics and photonics, terahertz radiation has many attractive properties and has great potential for applications in areas such as wireless communications [1–4], imaging [5–7], and biomedicine [8–10]. On the other hand, the terahertz band is known as the "terahertz gap" due to the inadequate power of sources.

As a popular type of vacuum electronic device, the TWT can amplify terahertz signals and give remarkable output power with much higher efficiency over solid-state power amplifiers, which is critical for the application of terahertz waves.

In recent years, many institutes have conducted experimental research on terahertz TWTs. Hu et al. [11,12] reported the development of a 0.22 THz and a 0.34 THz TWT at the China Academy of Engineering Physics. Their maximum output powers reach 30 W and 3.17 W, respectively. The corresponding gains are 31.2 dB and 26.2 dB, respectively. Liu et al. [13] of the Aerospace Information Research Institute reported a power amplifying scheme by two cascaded TWTs at 0.22 THz and achieved a 60 W peak output power and a 33 dB peak gain. The recent advances in terahertz TWTs at the Beijing Vacuum Electronic Research Institute were reported by Pan et al. [14,15]; they successfully fabricated an 11.9 W 25.5 dB 0.26 THz TWT and a 1.6 W 0.34 THz power module, in which a TWT with a maximum gain of 22 dB was applied. Northrop Grumman Corporation developed a series of high-frequency terahertz TWTs. The operating frequency band includes 0.67 THz [16],

0.85 THz [17], and 1.03 THz [18]. Among them, the 1.03 THz TWT holds the highest operating frequency record for TWTs to date, which has a peak output power of 29 mW and a peak gain of 20 dB.

It is known that fabrication tolerance is very crucial at the terahertz band; thus, a SWS with a simple structure is helpful for the success of a terahertz TWT. In our previous work, we proposed an attenuator-free SWS, named the modified angular log-periodic (MALP) FW [19], for single-section high-gain terahertz TWTs and validated its principle by a prototype TWT at the Ka band [20]. Based on the concept of the QSR in the MALP-FW SWS, this paper proposed a novel SQSST-FW SWS to further reduce the fabrication difficulty and improve the output power of terahertz TWTs. In addition, a prototype 650 GHz SQSST-FW TWT is designed in this paper as an example to illuminate the scheme and design method. According to the simulation, it can produce an output of over 3 W with an input of 0.4 mW around 650 GHz. As a high-power terahertz source, the SQSST-FW TWT would benefit the development of the terahertz sensors for fields including nondestructive inspection and testing, electromagnetic biology effects, and recognition of protein structural states.

The structure of this paper is as follows: Section 2 describes the scheme of the SQSST-FW SWS; Section 3 briefly reviews the small-signal theoretical foundation of the backward-wave oscillations in the SQSST-FW SWS; Section 4 verifies the accuracy of the small-signal theory in predicting the starting length of oscillation (hereafter referred to as starting length) by using particle-in-cell (PIC) simulations; Section 5 introduces the simulation for the performance of the designed 650 GHz TWT.

## 2. Scheme of SQSST-FW SWS

Figure 1 shows the overall structure of the SQSST-FW SWS, including 9 segments marked in different colors. The overall structure is symmetrical about the middle plane in the z-direction. The zoom-in view shows the line art of a unit cell, where the deep blue dash curve is the meandering path. The total length and the projected length in the z-axis of the meandering path are $L$ and $p$, respectively. Within one segment, there are several unit cells with the same $L$ and $p$, which are different in different segments.

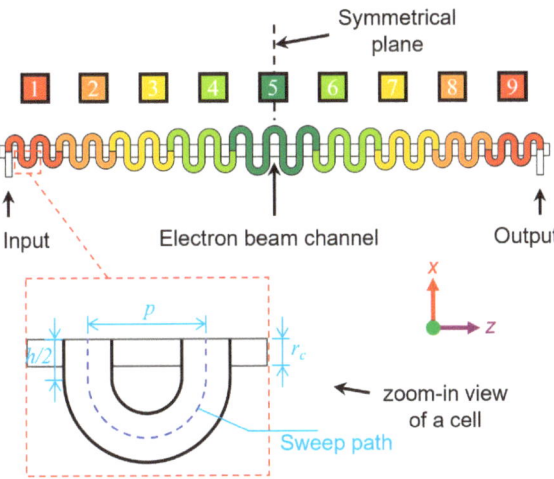

**Figure 1.** A schematic diagram of the SQSST-FW SWS.

The first half of the SQSST-FW SWS becomes wider gradually, being similar to that in [19]. By changing the values of $h$ and $p$, we made the SQSST-FW SWS have a QSR. The significant difference from the MALP-FW SWS is that the SQSST-FW SWS is step-transition by segments, while the latter is transition by every cell. The novel topology reduces the requirement for SWS machining accuracy. On the other hand, the performance of suppressing self-oscillations is also reduced. Hence, the number of cells in every segment should be limited to avoid oscillations, which will be discussed in Sections 3 and 4.

In addition, the symmetrical design makes the axial wave phase velocities vary in reversed trends in the two halves, allowing for the creation of a positive-/negative-tapering phase-velocity curve in a certain frequency range and the improvement of output power.

With waveguide cross-sectional dimensions of 0.26 mm × 0.055 mm and the electron beam channel radius $r_c$ of 0.05 mm, Table 1 lists the other dimensions of the cells in different segments of the 650 GHz SQSST-FW SWS, where the subscript represents the number of the segment.

**Table 1.** Cell dimensions in different segments.

| Symbol | Value (mm) | Symbol | Value (mm) |
|---|---|---|---|
| $p_1$ | 0.103 | $h_1$ | 0.105 |
| $p_2$ | 0.10403 | $h_2$ | 0.109 |
| $p_3$ | 0.10478 | $h_3$ | 0.113 |
| $p_4$ | 0.10578 | $h_4$ | 0.118 |
| $p_5$ | 0.10682 | $h_5$ | 0.123 |

As mentioned above, the number of cells in every segment should be limited due to the reduction in the oscillation-suppressing performance. A simple method to determine this number is based on the small-signal theory introduced in the following sections.

## 3. Analysis of the Oscillations

The most common self-oscillations in TWTs can be divided into reflection oscillation and backward-wave oscillation [21]. The reflections of the wave are the main contributor to the reflection oscillations because they construct the energy feedback loop. Fortunately, the oscillations caused by the reflections can be naturally suppressed in the terahertz TWTs due to the high transmission loss. Therefore, this section focuses on the analysis of backward-wave oscillations.

The small-signal equations of backward-wave oscillators (BWOs), whose detailed derivation was given in Liu's book [21], were employed to evaluate the effect of backward-wave oscillations in the TWT. The key equations are reorganized and summarized here.

### 3.1. 1-D Characteristic Equation for BWOs

In the small-signal theory, one-dimensional (1-D) electronic Equation (1), which describes the effect of the circuit field on the electron beam, is the same for a BWO and a TWT.

$$i_1 = \frac{j\beta_e}{(\Gamma - j\beta_e)^2 + \beta_q^2} \frac{I_0 E_c}{2V_0} \quad (1)$$

Here, $i_1$ is the AC component of the beam current; $\beta_e$ is the phase constant for the electron beam; $\Gamma$ is the propagation constant of the waves with the beam loaded (to be solved); $\beta_q$ is the phase constant of the plasma wave; $I_0$ is the dc beam current; $E_c$ is the circuit field; and $V_0$ is the dc beam voltage.

The circuit equation, which describes the effect of the electron beam on the circuit field, for BWOs is also essentially the same as that for TWTs except for the sign change caused by the change in the equation of the definition of the interaction impedance, as follows:

$$E_c = \frac{\Gamma^2 \Gamma_0 K_c}{\Gamma^2 - \Gamma_0^2} i_1 \quad (2)$$

where $\Gamma_0$ is the propagation constant of the wave in the SWS without the electron beam loaded; $K_c$ is the interaction impedance.

Joining (1) and (2), the characteristic equation for BWOs can be obtained, as follows:

$$\left(\frac{\Gamma^2 \Gamma_0 K_c}{\Gamma^2 - \Gamma_0^2}\right)\left[\frac{j\beta_e}{(\Gamma - j\beta_e)^2 + \beta_q^2}\frac{I_0}{2V_0}\right] - 1 = 0 \qquad (3)$$

Following the method for solving the characteristic equation for TWTs, let

$$\begin{cases} \Gamma = j\beta_e - \beta_e C\delta \\ \Gamma_0 = j\beta_e + j\beta_e Cb - \beta_e Cd \end{cases} \qquad (4)$$

and

$$\delta = x + jy. \qquad (5)$$

where C, b, and d are the gain, velocity, and loss parameters, respectively.

By substituting (4) and (5) into (3) and assuming that $C << 1$, and that b, d, and $\delta$ are not far from 1, one can obtain a pair of numerically solvable equations of x and y, as follows:

$$\begin{cases} (x^2 - y^2 + 4QC)(d - x) + 2xy(y + b) = 0 \\ (x^2 - y^2 + 4QC)(b + y) + 2xy(x - d) - 1 = 0 \end{cases} \qquad (6)$$

### 3.2. The Starting Length of the Backward-Wave Oscillation

Similar to the case in TWTs, the gain equation for BWOs can be obtained by associating the boundary conditions, the electronic equation, and the current continuity equation. The gain of the backward wave "G" is

$$G = \frac{\sum_{n=1}^{3}(\delta_{n+2} - \delta_{n+1})(\delta_n^2 + 4QC)e^{j2\pi N}}{\sum_{n=1}^{3}(\delta_{n+2} - \delta_{n+1})(\delta_n^2 + 4QC)e^{j2\pi CN\delta_n}}. \qquad (7)$$

where $\delta_5 = \delta_2$; $\delta_4 = \delta_1$; QC is the space charge parameter; and N is the ratio of the circuit length to the wavelength for the electron beam.

Apparently, the starting condition of the backward-wave oscillation is $G \to \infty$ or $1/G = 0$, that is,

$$\sum_{n=1}^{3}\frac{(\delta_n^2 + 4QC)e^{j2\pi CN\delta_n}}{(\delta_{n+2} - \delta_n)(\delta_{n+1} - \delta_n)} = 0. \qquad (8)$$

$\delta$, Q, and C are all functions of the cell dimensions and the beam parameters, so the starting length of the backward-wave oscillation can be obtained by finding the proper b and CN that satisfy (8).

## 4. Starting Length in a Periodic SWS

### 4.1. Illumination of the Sample SWS

Here, a periodic FW SWS model was built and simulated for Segment 1 using PIC methods in CST [22] Particle Studio to verify the accuracy of the predicted starting length from (8).

The simulation results of the dispersion characteristic curves of this periodic SWS are shown in Figure 2a, from which the frequency of the backward-wave oscillation point is found to be around 850 GHz, and the interaction impedance around the backward-wave oscillation point is shown by Figure 2b, where r represents the radial coordinate.

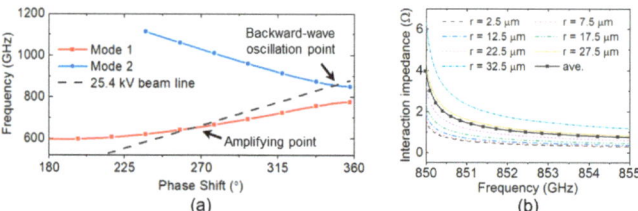

**Figure 2.** (**a**) A dispersion diagram and (**b**) the interaction impedance around the backward-wave oscillation point of the sample for periodic FW SWS.

To obtain the accurate oscillation frequency, a zero-drive PIC simulation model with an electron beam of 25.4 kV and 15 mA was built, and the simulation results found it to be 851 GHz, where the beam radius is 0.035 mm, and a 1 T uniform focusing magnetic field is applied. The average beam–wave interaction impedance of the backward wave at that point obtained by numerical simulation was 1.5 Ω.

### 4.2. Numerical Solutions of the Characteristic Equation

A prerequisite of numerically solving (6) is that $Q$, $C$, $b$, and $d$ are known, where $Q$ and $C$ can be directly obtained by the known beam parameter and the interaction impedance. The loss parameter $d$ is not directly calculable but can be obtained by substituting the simulation result of insertion loss into its definition equation. When neglecting the conductor loss of the SWS material, $d$ is equal to zero apparently, while it changes to 0.507 if a conductivity of $2 \times 10^7$ S/m was considered. The numerical solutions of (6) under these two conditions are given in Figure 3.

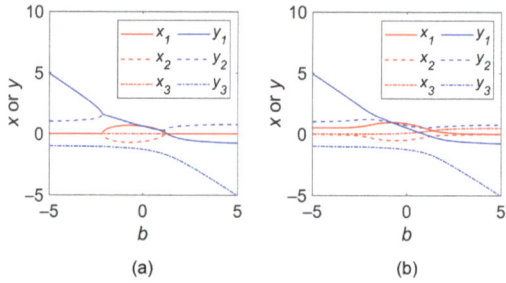

**Figure 3.** Solutions of $\delta$: (**a**) $d = 0$; and (**b**) $d = 0.507$.

### 4.3. Calculated Starting Length Using the Small-Signal Equation

Once $\delta$ is solved, $G$ or $1/G$ would be a singular value function of $CN$. The variation of $1/G$ with $CN$ for different values of $b$ is shown in Figure 4.

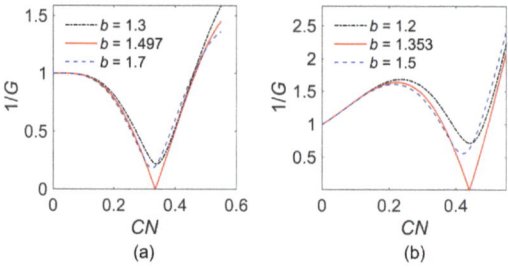

**Figure 4.** Variation of $1/G$ with $CN$: (**a**) $d = 0$; and (**b**) $d = 0.507$.

The red solid curve in Figure 4a indicates that the starting condition of the backward-wave oscillation in the loss-free model of the sample SWS is $b = 1.497$ and $CN > 0.336$, which corresponds to an actual starting length of about 5.95 mm. Also, the starting length under the lossy condition ($\sigma = 2 \times 10^7$ S/m) shown in Figure 4b is about 7.79 mm.

## 4.4. Starting Length by PIC Simulation

In the loss-free condition, the time-domain signal output from the "input port" of the TWT is shown in Figure 5. The number of cells was gradually added until a remarkable oscillation was observed. In order to characterize the oscillation more clearly, the time-dependent spectra were computed by Fast Fourier Transform (FFT) with a 0.2 ns rectangular window, which is shown in Figure 6.

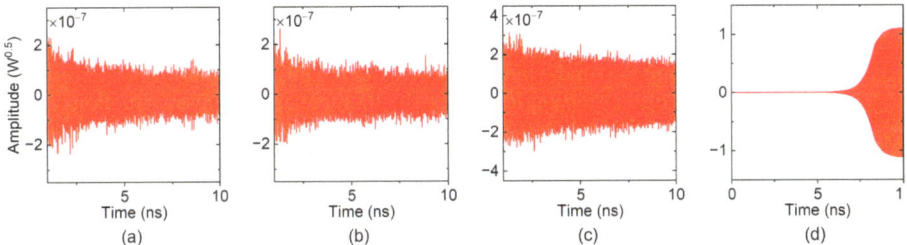

**Figure 5.** Time-domain signals from the "input port" of the loss-free sample TWT: (**a**) 52 cells; (**b**) 56 cells; (**c**) 60 cells; and (**d**) 62 cells.

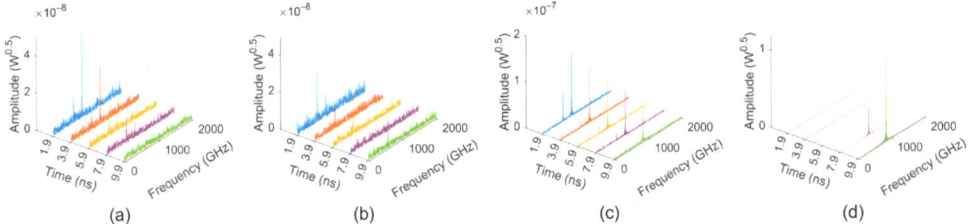

**Figure 6.** Frequency spectra of the signals from the "input port" of the loss-free sample TWT: (**a**) 52 cells; (**b**) 56 cells; (**c**) 60 cells; and (**d**) 62 cells.

The frequency spectra of the 52- and 56-cell TWT look disordered, and their magnitudes are at the level of $10^{-8}$–$10^{-7}$, as shown in Figure 6a,b. This means that there is no obvious oscillation formed in these situations. As the number of cells increased to 60, a clear peak occurred in Figure 6c, whose frequency gradually moved to 851 GHz, with a main backward-wave oscillation frequency of mode 2. However, its magnitude is rather very low ($10^{-7}$ level), and it is stable and does not grow exponentially over time, which means oscillations are primed but not pronounced. When the number of cells is further increased to 62, a typical self-oscillation signal arises, as shown in Figure 5d. These results indicate that the starting number of cells for the loss-free sample SWS is 56–62, corresponding to a starting length of 5.77–6.39 mm.

Figure 7 shows the frequency spectrum results in the lossy situation, indicating that the starting number of cells for the lossy sample SWS is 72–76, corresponding to a starting length of 7.42–7.83 mm.

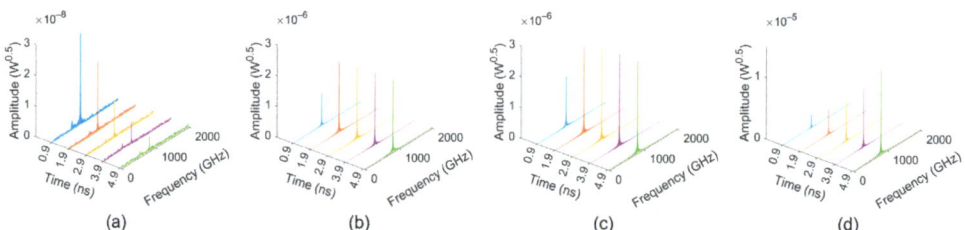

**Figure 7.** Frequency spectrum of the signals from the "input port" of the lossy sample TWT: (**a**) 60 cells; (**b**) 72 cells; (**c**) 74 cells; and (**d**) 76 cells.

*4.5. Comparison between Analytical and Simulation Solutions*

This section compares the starting length for a periodic FW SWS by using two methods, which are the small-signal equation and the PIC simulation, respectively. The calculated results of the starting length for the loss-free and the lossy situations are 5.95 mm and 7.79 mm, respectively, using the small-signal equation. The computed results by the PIC simulation are in two ranges, namely 5.77–6.39 mm and 7.42–7.83 mm. The results provided by the two methods show good agreement. Hence, the simple small-signal equation would be a convenient tool for calculating the starting length in a periodic FW SWS with great accuracy.

## 5. Performance Simulation

*5.1. Simulation of the Dispersion Characteristics*

Figure 8 shows the simulation results of the cell dispersion characteristics in different segments, where the part of Segments 6–9 was neglected due to the symmetry of the structure.

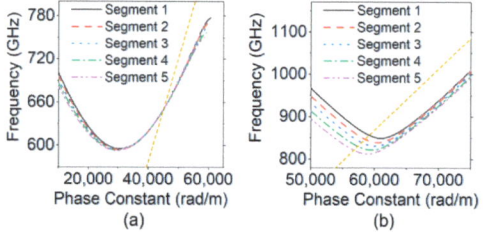

**Figure 8.** Simulation results of the dispersion characteristics of designed 650 GHz SQSST-FW SWS: (**a**) mode 1; and (**b**) mode 2.

The frequency of the designed perfect-synchronous point (PSP) is 650 GHz, which can be observed in Figure 8a. By the same method in Section 4, one can find that the oscillation frequencies in Segments 2–5 are 841.8 GHz, 833.9 GHz, 824.5 GHz, and 816.2 GHz, respectively, corresponding to the main backward-wave oscillation of mode 2, as shown in Figure 8b.

*5.2. Starting Lengths in Segments 2–5*

Table 2 lists the parameters of these oscillation points, where the conductivity of the background metal is still set as $\sigma = 2 \times 10^7$ S/m.

**Table 2.** Oscillation-point parameters in Segments 2–5.

| Segment | Frequency (GHz) | Interaction Impedance (Ω) | Loss Parameter $d$ |
|---|---|---|---|
| 2 | 841.8 | 1.5 | 0.511 |
| 3 | 833.9 | 1.6 | 0.516 |
| 4 | 824.5 | 1.8 | 0.489 |
| 5 | 816.2 | 1.9 | 0.473 |

The diagrams of $1/G$ versus $CN$ for Segments 2–5 can then be easily obtained by substituting the data in Table 2 into (6), as shown in Figure 9.

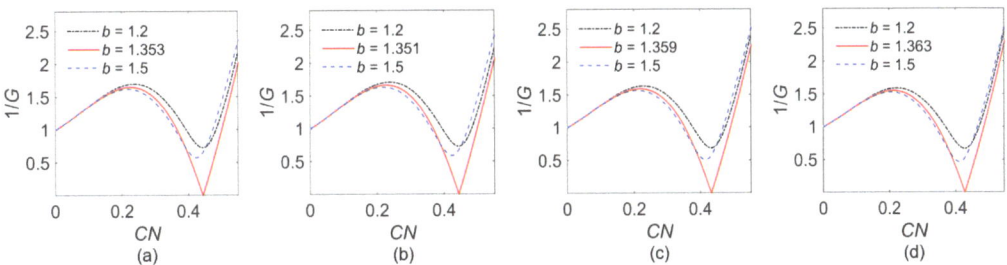

**Figure 9.** Relations of $1/G$ versus $CN$: (**a**) Segment 2; (**b**) Segment 3; (**c**) Segment 4; and (**d**) Segment 5.

The calculated results in Figure 9 indicate that the starting lengths for Segments 2–5 are 7.91 mm, 7.82 mm, 7.41 mm, and 7.27 mm, respectively, corresponding to cell numbers 76, 74, 70, and 68.

*5.3. Overall Structure Design and Transmission Characteristics*

As the oscillation frequencies for Segments 1–5 are different, the potential backward-wave oscillations in them are non-coherent. So, the oscillation would not start as long as the number of cells with different dimensions in the overall SWS remains less than the starting number of cells.

Following the above principle, the number of cells in Segment 5 is designed as a conservative value of 63, and that in others are all 33.

The simulation results of the transmission characteristics of the overall SWS are shown in Figure 10, where $S_{11}$ is about $-15$ dB to $-17$ dB, and $S_{21}$ is about $-45$ dB to $-65$ dB in the frequency range of 630 GHz–670 GHz.

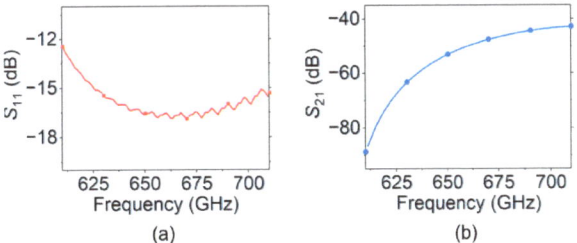

**Figure 10.** Simulation results of the characteristics of the 650 GHz SQSST-FW SWS: (**a**) $S_{11}$; and (**b**) $S_{21}$.

*5.4. PIC Simulation*

A 25.4 kV 15 mA ideal electron beam with a radius of 0.035 mm was then applied to the PIC simulation. At first, a zero-drive simulation was performed to evaluate the oscillation state in the designed 650 GHz SQSST-FW TWT. Figure 11 shows the time-domain output signals from the two ports of the TWT.

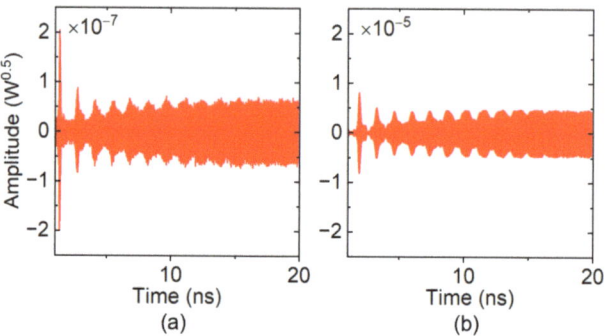

**Figure 11.** Time-domain signals from the (**a**) "input port" and (**b**) "output port" of the 650 GHz SQSST-FW TWT.

The simulation results in Figure 11 indicate that the powers of the self-excited signals from the two ports of the TWT are quite low, whereas the power output from the "output port" is at the pW level and that from the "input port" is even at the fW level. The insets in Figure 11a,b show that the self-excited signals have no upward trend. The frequency spectra of these two signals are shown in Figure 12, in which one can find that the peak frequency is very close to 650 GHz, the frequency of the PSP.

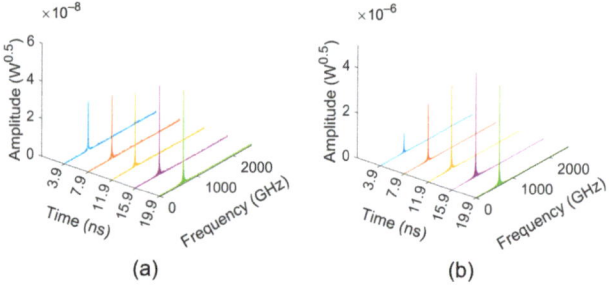

**Figure 12.** Frequency spectrum of the (**a**) "input port" and (**b**) "output port" signals of the 650 GHz SQSST-FW TWT.

According to the above information, it is reasonable to determine the output signal from the "output port" of the TWT as a tiny reflection oscillation, and the output signal from the "input port" as its reflection. The reason for such a tiny oscillation is that the phase velocity of the forward wave of 650 GHz, the frequency of the PSP, is constant along the $z$-axis, and thus the SWS has no function of suppressing reflection oscillation of this frequency.

However, there is no positive energy feedback loop for the reflection oscillation in the tube, and the steady oscillation power is only at the pW level. So, the effect of this tiny oscillation on the performance of the tube can be neglected.

Figure 13 shows the drive curves in the frequency range of 641 GHz–651 GHz with a step of 1 GHz, and Figure 14 shows the output powers at different frequencies with a fixed 0.4 mW input power.

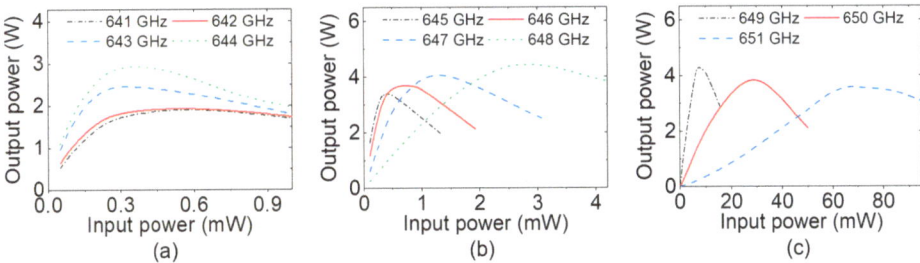

**Figure 13.** PIC simulation results of the drive curves of the 650 GHz SQSST-FW TWT: (**a**) 641 GHz–644 GHz; (**b**) 645 GHz–648 GHz; and (**c**) 649 GHz–651 GHz.

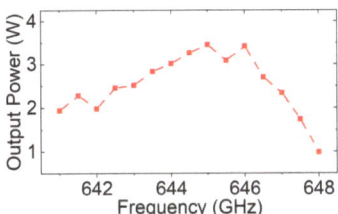

**Figure 14.** PIC simulation results of the equal-drive output powers at different frequencies.

The PIC simulation results indicate that the designed SQSST-FW TWT has a saturated output power of 3.46 W when the input power is 0.43 mW at 645 GHz, corresponding to a gain of 39.1 dB and an electronic efficiency of 0.91%. In addition, the maximum output power can reach 4.46 W when the input power is 2.8 mW at 648 GHz. The gain would be reduced to 32 dB, but the electronic efficiency would increase to 1.17%. In addition, the 3 dB bandwidth with a 0.4 mW input power is about 6.5 GHz.

## 6. Discussion

In Sections 3 and 4, the backward-wave oscillation starting length for the sample periodic FW SWS is analyzed by 1-D linear theory for BWOs and successfully verified by PIC simulations. However, it has actually been conditionally simplified there, including ignoring the impact of the plasma frequency reduction factor and adopting the simplified formula for the space charge parameter ($QC = \omega_p^2/\omega^2/C^2/4$, where $\omega_p$ is the plasma frequency). The condition is that both $C$ and $QC$ are relatively low, which can be satisfied in most terahertz FW TWTs. For the TWTs with high $C$ or high $QC$, the calculation results may need to be modified.

The designed 650 GHz SQSST-FW TWT performs well in gain, output power and electronic efficiency but shows disadvantages in some other respects. One disadvantage is its relatively narrow 3 dB bandwidth, about 1%, and another is the high requirement for machining accuracy. Further detailed studies need to be conducted to explore its optimization potential.

## 7. Conclusions

This paper presents a novel SQSST-FW SWS based on the QSR and a design method for it by the combination of the 1-D linear theory for BWOs and numerical simulations, which is illuminated by a design example for 650 GHz TWT applications. The designed SQSST-FW SWS consists of nine segments, which are symmetrical about the fifth segment. Following the principle that the cell number in every segment did not exceed the oscillation-starting cell number, the finally designed 650 GHz SQSST-FW TWT reached a maximum saturated gain of 39.1 dB without the help of attenuators or severs in PIC simulation. It also performs well in output power and electronic efficiency. The simulation results verify the validity of the proposed concept and method and show the great potential of the SQSST-FW SWS for applications in terahertz high-gain high-power TWTs.

**Author Contributions:** Conceptualization, D.X. and S.W.; methodology, D.X. and T.H.; software, D.X. and T.H.; validation, Y.Z., Z.L. and Z.W.; formal analysis, D.X. and H.G.; investigation, D.X. and Z.W.; resources, Z.W., Z.D. and S.W.; data curation, T.H. and Z.D.; writing—original draft preparation, D.X.; writing—review and editing, Z.D. and S.W.; visualization, Z.L.; supervision, S.W.; project administration, S.W.; funding acquisition, S.W. All authors have read and agreed to the published version of the manuscript.

**Funding:** This research was funded by the National Natural Science Foundation of China (grant nos. T2241002, 61921002, 92163204, and 61988102) and the National Key Laboratory of Science and Technology on Vacuum Electronics (grant no. 2023KP006).

**Institutional Review Board Statement:** Not applicable.

**Informed Consent Statement:** Not applicable.

**Data Availability Statement:** Data are contained within the article.

**Conflicts of Interest:** The authors declare no conflicts of interest.

# References

1. Kanno, A.; Sekine, N.; Kasamatsu, A.; Yamamoto, N.; Yoshida, M.; Masuda, N. Terahertz-Wave Communication System Using a Traveling-Wave Tube Amplifier. In Proceedings of the 2018 Progress in Electromagnetics Research Symposium (PIERS-Toyama), Toyama, Japan, 1–4 August 2018.
2. Lin, C.; Li, G.Y.L. Terahertz Communications: An Array-of-Subarrays Solution. *IEEE Commun. Mag.* **2016**, *54*, 124–131. [CrossRef]
3. Song, R.; Cui, D.; Li, Y.; Zhang, N.; Lv, Q.; Wang, C. The Progress of Terahertz Wave Source in Communication. In Proceedings of the 2016 IEEE 9th UK-Europe-China Workshop on Millimetre Waves and Terahertz Technologies (UCMMT), Qingdao, China, 5–7 September 2016; pp. 64–66.
4. Zhou, T.; Zhang, Y.; Zhang, B.; Zeng, H.; Tan, Z.; Zhang, X.; Wang, L.; Chen, Z.; Cao, J.; Song, K.; et al. Terahertz Direct Modulation Techniques for High-Speed Communication Systems. *China Commun.* **2021**, *18*, 221–244. [CrossRef]
5. Hu, Q.; Wei, X.; Pang, Y.; Lang, L. Advances on Terahertz Single-Pixel Imaging. *Front. Phys.* **2022**, *10*, 982640. [CrossRef]
6. She, R.; Liu, W.; Lu, Y.; Zhou, Z.; Li, G. Fourier Single-Pixel Imaging in the Terahertz Regime. *Appl. Phys. Lett.* **2019**, *115*, 021101. [CrossRef]
7. Shen, Y.C.; Gan, L.; Stringer, M.; Burnett, A.; Tych, K.; Shen, H.; Cunningham, J.E.; Parrott, E.P.J.; Zeitler, J.A.; Gladden, L.F.; et al. Terahertz Pulsed Spectroscopic Imaging Using Optimized Binary Masks. *Appl. Phys. Lett.* **2009**, *95*, 231112. [CrossRef]
8. Jiao, Y.; Lou, J.; Ma, Z.; Cong, L.; Xu, X.; Zhang, B.; Li, D.; Yu, Y.; Sun, W.; Yan, Y.; et al. Photoactive Terahertz Metasurfaces for Ultrafast Switchable Sensing of Colorectal Cells. *Mater. Horiz.* **2022**, *9*, 2984–2992. [CrossRef] [PubMed]
9. Wang, P.; Lou, J.; Yu, Y.; Sun, L.; Sun, L.; Fang, G.; Chang, C. An Ultra-Sensitive Metasurface Biosensor for Instant Cancer Detection Based on Terahertz Spectra. *Nano Res.* **2023**, *16*, 7304–7311. [CrossRef]
10. Wang, R.; Xu, L.; Huang, L.; Zhang, X.; Ruan, H.; Yang, X.; Lou, J.; Chang, C.; Du, X. Ultrasensitive Terahertz Biodetection Enabled by Quasi-Bic-Based Metasensors. *Small* **2023**, *19*, 2301165. [CrossRef] [PubMed]
11. Hu, P.; Lei, W.; Jiang, Y.; Huang, Y.; Song, R.; Chen, H.; Dong, Y. Development of a 0.32-THz Folded Waveguide Traveling Wave Tube. *IEEE Trans. Electron Devices* **2018**, *65*, 2164–2169. [CrossRef]
12. Hu, P.; Lei, W.; Jiang, Y.; Huang, Y.; Song, R.; Chen, H.; Dong, Y. Demonstration of a Watt-Level Traveling Wave Tube Amplifier Operating above 0.3 THz. *IEEE Electron Device Lett.* **2019**, *40*, 973–976. [CrossRef]
13. Liu, W.; Zhang, L.; Liu, W.; Jin, Z.; Zhao, K.; Zhao, C.; Guo, X.; Zhang, Z.; Yin, S.; Zhang, Z. Demonstration of a High-Power and Wide-Bandwidth G-Band Traveling Wave Tube with Cascade Amplification. *IEEE Electron Device Lett.* **2021**, *42*, 593–596. [CrossRef]
14. Pan, P.; Zhang, L.; Cui, H.; Feng, J. Terahertz Power Module Based on 0.34 THz Traveling Wave Tube. *IEEE Electron Device Lett.* **2022**, *43*, 816–819. [CrossRef]
15. Pan, P.; Zheng, Y.; Li, Y.; Song, X.; Feng, Z.; Feng, J.; Britt, R.D.; Luhmann, N.C. Demonstration of a 263-GHz Traveling Wave Tube for Electron Paramagnetic Resonance Spectroscopy. *IEEE Trans. Electron Devices* **2023**, *70*, 5897–5902. [CrossRef] [PubMed]
16. Tucek, J.C.; Basten, M.A.; Gallagher, D.A.; Kreischer, K.E.; Lai, R.; Radisic, V.; Leong, K.; Mihailovich, R. A 100 Mw, 0.670 THz Power Module. In Proceedings of the International Conference on Vacuum Electronics, Monterey, CA, USA, 24–26 April 2012.
17. Tucek, J.C.; Basten, M.A.; Gallagher, D.A.; Kreischer, K.E. 0.850 THz Vacuum Electronic Power Amplifier. In Proceedings of the IEEE International Vacuum Electronics Conference, Monterey, CA, USA, 22–24 April 2014.
18. Tucek, J.C.; Basten, M.A.; Gallagher, D.A.; Kreischer, K.E. Operation of a Compact 1.03 THz Power Amplifier. In Proceedings of the 2016 IEEE International Vacuum Electronics Conference (IVEC), Monterey, CA, USA, 19–21 April 2016.

19. Xu, D.; Wang, S.; Wang, Z.; Shao, W.; He, T.; Wang, H.; Tang, T.; Gong, H.; Lu, Z.; Duan, Z.; et al. Theory and Experiment of High-Gain Modified Angular Log-Periodic Folded Waveguide Slow Wave Structure. *IEEE Electron Device Lett.* **2020**, *41*, 1237–1240. [CrossRef]
20. Xu, D.; Wang, S.; Lu, C.; He, T.; Wang, Z.; Lu, Z.; Gong, H.; Duan, Z.; Gong, Y. Demonstration of a Modified Angular Log-Periodic Folded Waveguide Traveling Wave Tube at Ka-Band. *IEEE Trans. Electron Devices* **2023**, *70*, 1323–1329. [CrossRef]
21. Liu, S. *Introduction to Microwave Electronics*; National Defense Industry Press: Beijing, China, 1985. (In Chinese)
22. Cst Studio Suite. Available online: https://www.3ds.com/products-services/simulia/products/cst-studio-suite (accessed on 21 November 2022).

**Disclaimer/Publisher's Note:** The statements, opinions and data contained in all publications are solely those of the individual author(s) and contributor(s) and not of MDPI and/or the editor(s). MDPI and/or the editor(s) disclaim responsibility for any injury to people or property resulting from any ideas, methods, instructions or products referred to in the content.

*Communication*

# A Staggered Vane-Shaped Slot-Line Slow-Wave Structure for W-Band Dual-Sheet Electron-Beam-Traveling Wave Tubes

Yuxin Wang [1], Jingyu Guo [1], Yang Dong [1], Duo Xu [1], Yuan Zheng [1], Zhigang Lu [1], Zhanliang Wang [1] and Shaomeng Wang [1,2,*]

[1] School of Electronic Science and Engineering, University of Electronic Science and Technology of China, No. 2006 Xiyuan Avenue, High-Tech District (West District), Chengdu 611731, China; 202211022437@std.uestc.edu.cn (Y.W.); 202111022436@std.uestc.edu.cn (J.G.); 202111022405@std.uestc.edu.cn (Y.D.); d.xu@uestc.edu.cn (D.X.); zyzheng@uestc.edu.cn (Y.Z.); lzhgchnn@uestc.edu.cn (Z.L.); wangzl@uestc.edu.cn (Z.W.)

[2] Yangtz Delta Region Institute (Quzhou), University of Electronic Science and Technology of China, No. 1, Chengdian Road, Quzhou 324003, China

\* Correspondence: wangsm@uestc.edu.cn

**Abstract:** A staggered vane-shaped slot-line slow-wave structure (SV-SL SWS) for application in W-band traveling wave tubes (TWTs) is proposed in this article. In contrast to the conventional slot-line SWSs with dielectric substrates, the proposed SWS consists only of a thin metal sheet inscribed with periodic grooves and two half-metal enclosures, which means it can be easily manufactured and assembled and has the potential for mass production. This SWS not only solves the problem of the dielectric loading effect but also improves the heat dissipation capability of such structures. Meanwhile, the SWS design presented here covers a $-15$ dB $S_{11}$ frequency range from 87.5 to 95 GHz. The 3-D simulation for a TWT based on the suggested SWS is also investigated. Under dual-electron injection conditions with a total voltage of 17.2 kV and a total current of 0.3 A, the maximum output power at 90 GHz is 200 W, with a 3 dB bandwidth up to 4 GHz. With a good potential for fabrication using microfabrication techniques, this structure can be a good candidate for millimeter-wave TWT applications.

**Keywords:** slot line; slow-wave structure (SWS); W-band; dual sheet electron beam; traveling wave tube

## 1. Introduction

Miniaturized vacuum electronics are commonly employed in millimeter-wave bands for applications such as high-data-rate wireless communications, satellite communications, and high-resolution radars [1,2]. Among them, the planar traveling wave tube (TWT) [3–5] is one of the most promising vacuum electronic devices due to its simple structure and compatibility with microfabrication methods in the millimeter-wave band.

One core part of a TWT is its slow-wave structure (SWS), which serves to slow down and synchronize the electromagnetic waves with a high-energy electron beam. Meander-line SWSs [6–9] have been extensively utilized in planar SWSs due to their low operating voltage and mass manufacturing capability. However, meander-line (ML) SWSs are typically kept in place by dielectric rods or dielectric substrates, which are susceptible to charge accumulation effects that can cause destructive damage to the device [10,11].

Dielectric slab-supported slot-line SWS [12] and co-planar SWS [13] can effectively minimize the dielectric's exposed area, thereby reducing the probability of charges bombarding the dielectric. The backward wave has a higher coupling impedance; hence, this SWS is commonly employed in backward wave oscillators (BWOs). However, the presence of the dielectric limits the thickness of the metal layer in these two types of SWSs, and in actual processing, the loss of the intermediate seed layer, which connects the dielectric to the metal layer, can have a negative impact on wave transmission [14].

Due to the size co-transition effect, the transverse dimension of the SWS decreases with increasing frequency, and then the use of large compression ratios for multiple sheet electron beam injections can effectively increase the output power. Several designs of multiple-tunnel TWTs operating in millimeter-wave bands have been proposed [15,16].

In this condition, a novel planar-staggered vane-shaped slot-line (SV-SL) SWS is proposed. This slot-line SWS dispenses with a dielectric substrate and consists of a metal sheet and a metal shell. The structure operates in the traveling wave region by adjusting the structural parameters. Similar to the meander-line SWS, the SV-SL SWS is simple to fabricate, can be mass-produced, and has better heat dissipation characteristics. Meanwhile, this SWS has two natural electron injection channels, which can be used for dual-electron injection operations and can effectively increase the output power.

## 2. SWS Design and Discussion
### 2.1. Design and Electromagnetic Parameters

The schematic design of the proposed SV-SL SWS with natural dual beam tunnels is shown in Figure 1. This model is partitioned into the following three parts: the upper shell, the middle sheet, and the lower shell. The distance between the upper and lower shells is drawn out for easy observation. The center part is engraved with staggered vane patterns and fixed between the grooved upper and lower shells, while the gaps between the shell and centerpiece become nature beam tunnels. The dimensions of the structure are presented in Table 1.

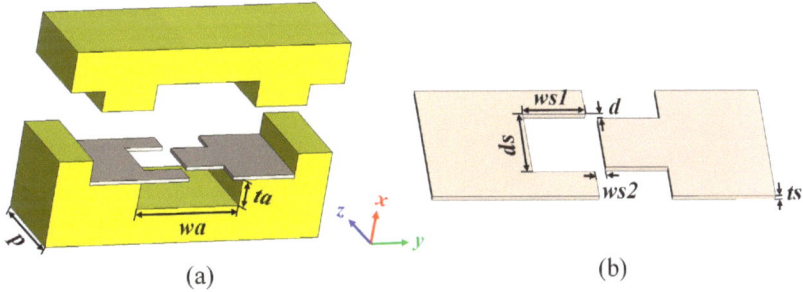

**Figure 1.** Schematic of the single period SV-SL SWS. (**a**) Perspective view of the proposed SWS with cut shell and (**b**) cross-section view of staggered vane slot line.

**Table 1.** Dimension parameters.

| Parameters | Dimension (mm) | Parameters | Dimension (mm) |
| --- | --- | --- | --- |
| $p$ | 1.2 | $ws1$ | 0.55 |
| $wa$ | 1.5 | $ws2$ | 0.1 |
| $ta$ | 0.375 | $d$ | 0.05 |
| $ds$ | 0.65 | $ts$ | 0.05 |

We simulated a single period of the proposed SWS with Floque periodic boundary conditions along the axial direction. The results are presented in Figure 2. To effectively avoid the occurrence of backward wave oscillations and band-edge oscillations, the electron injection voltage line should avoid intersecting the backward wave region (with phase shifts ranging from 0° to 180°) and move away from points with phase shifts of $\pi$ and $2\pi$. Therefore, by increasing the length of the single period and decreasing the height of the electron injection tunnels, the 17.2 kV beam line and the fundamental mode of the SWS intersect in the forward wave region at 90 GHz, which means that signals near this frequency may be amplified.

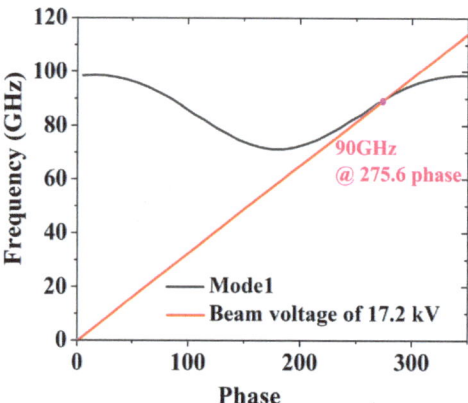

**Figure 2.** The dispersion diagram of the proposed SWS.

In addition, the variation in the normalized phase velocity and coupling impedance with frequency is shown in Figure 3. In the frequency range of 85–95 GHz, the dispersion characteristic curves are relatively flat, and the coupling impedance calculated for the fundamental forward harmonic mode of the SV-SL SWS is about 2.89–5.93 Ω at 0.075 mm from the surface of the slot line.

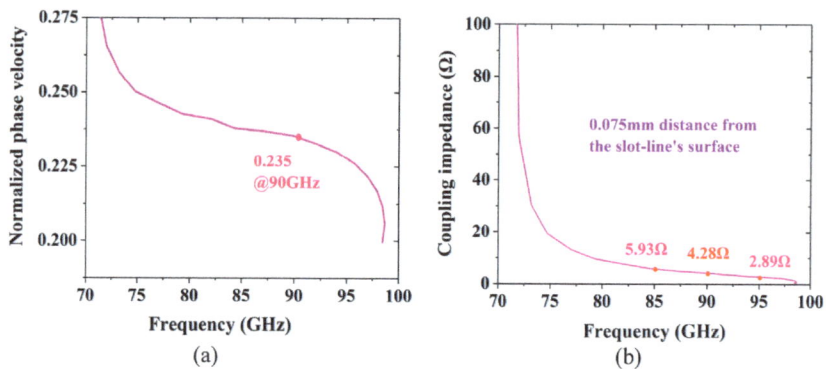

**Figure 3.** (**a**) Normalized phase velocity curve and (**b**) coupling impedance curve of the SV-SL SWS with frequency.

## 2.2. S-Parameters

Figure 4 depicts the whole assembly model of the proposed SWS with coupling devices. A thin central sheet of metal is laser-engraved with periodic staggered vane grooves, and the metal enclosure is divided into two halves that can be fabricated with a computer numerical control (CNC) milling or casting machine. The gap between the middle metal sheet and the upper and lower enclosures can serve as a dual electron injection channel.

To eliminate extra reflections, a transition section, as seen in Figure 5, is employed to connect the SWS to the input/output structure. A stepped ridge waveguide is usually adopted as the transmission transition structure for a planar meander-line SWS, where impedance matching can be achieved by adjusting the height of the ridge waveguide (the dimension along the x-direction); during this process, the transition from the $TE_{10}$ mode of the standard rectangular waveguide to the quasi-TEM mode of the meander line can be realized. However, this matching method requires the meander-line and ridge waveguide to be processed separately and then assembled as a whole unit, which is troublesome to accurately assemble and prone to introduce more assembly errors. In order to simplify the assembly of the whole SWS, in this SV-SL SWS, we controlled the height of the ridge waveguide in the x-direction to match the thickness of the centerpiece, i.e., the beam tunnel height, and achieved impedance matching by adjusting the dimensions of the ridge waveguide in the z- and y-directions. The dimensional parameters are listed in Table 2.

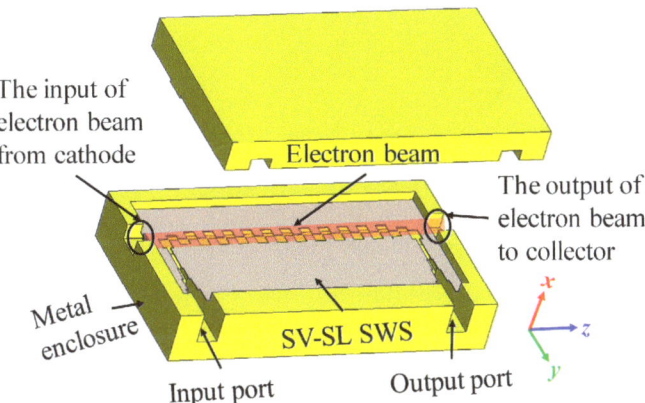

**Figure 4.** Assembly model of the proposed SWS with coupling structures.

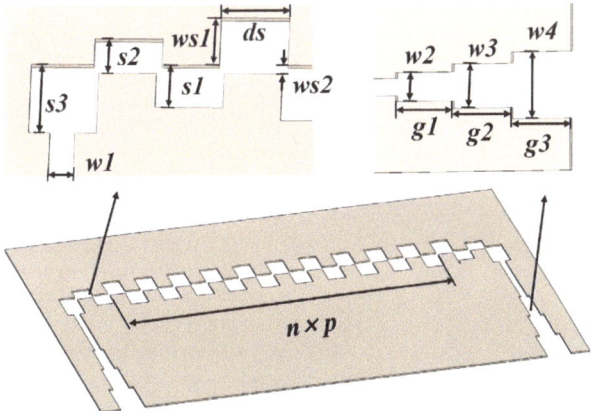

**Figure 5.** Schematic of the SV-SL SWS with transition coupler parts.

**Table 2.** Dimension parameters for transition.

| Parameters | Dimension (mm) | Parameters | Dimension (mm) |
|---|---|---|---|
| s1 | 0.5 | w3 | 0.6 |
| s2 | 0.4 | w4 | 0.9 |
| s3 | 0.8 | g1 | 0.95 |
| w1 | 0.24 | g2 | 1 |
| w2 | 0.39 | g3 | 1 |

Meanwhile, the electric field distribution of the transmission model of the proposed SWS is shown in Figure 6. After optimizing the dimensional parameters of the connection portion, this model can complete the wave transmission.

To examine the transmission characteristics of the SWS, we simulated a 50-period circuit with input/output couplers on both sides using the CST Studio Suite simulator [17].

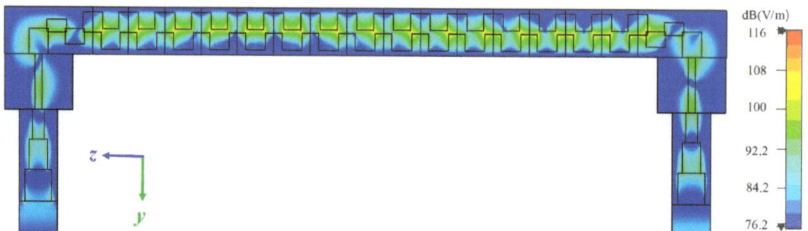

**Figure 6.** Electric field distribution of the transmission model of the proposed SWS.

In our previous research [18], the effects of laser cutting, acid cleaning, and copper plating on the surface roughness and S-parameter of the meander slot-line were investigated. The experimental results show that after plating a 3–5 µm copper layer on the surface of the laser-cut molybdenum sheet, the surface roughness reaches 1.67 µm, and its effective conductivity is $1.48 \times 10^7$ S/m. Therefore, in the SV-SL SWS simulation, the upper and lower shells are assumed to be copper with a conductivity of $3 \times 10^7$ S/m. And the central slot-line is considered to be a molybdenum sheet with an effective conductivity of $1.48 \times 10^7$ S/m; the effective conductivity $\sigma_{ef}$ was calculated according to the well-known formula.

$$\delta = \sqrt{\frac{2}{\omega\mu\sigma}} \tag{1}$$

$$\sigma_{ef} = \frac{\sigma}{(1 + \frac{2}{\pi}\arctan(1.4 \times (\frac{R}{\delta})^2))} \tag{2}$$

where the $\delta$ is the skin depth, $\omega$ is the angular frequency, $\mu = 4\pi \times 10^{-7}$ H/m is the magnetic conductivity, $\sigma = 5.8 \times 10^7$ S/m is the bulk conductivity of copper, and $R$ is the surface roughness.

The obtained S-parameters are presented in Figure 7a. In the frequency range of 87.5–95 GHz, the reflection loss $S_{11}$ does not exceed $-15$ dB (black line), and the transmission loss $S_{21}$ varies from $-5.5$ dB to $-8$ dB (red line). Due to the influence of high-frequency loss, the value of $S_{21}$ decreases significantly with frequency (the value of $S_{21}$ is negative while the absolute value of $S_{21}$ increases). During the actual processing, the meander line SWS with the dielectric support or substrate needs to consider the loss of the intermediate seed layer connecting the dielectric layer and the metal layer, which causes a serious negative impact on wave transmission [19,20]. Clearly, the transmission loss of SV-SL SWS with the seed layer removed is much lower.

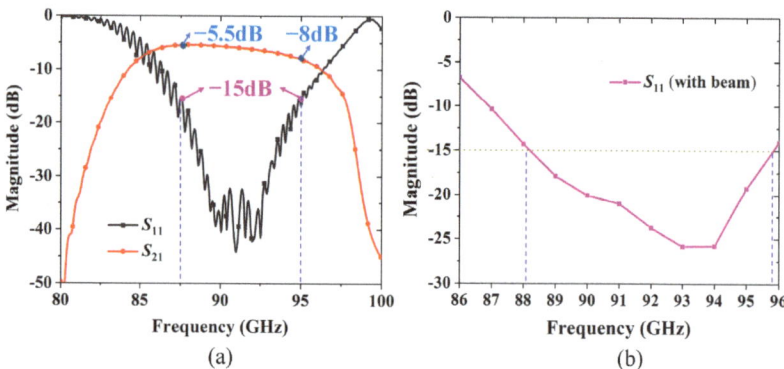

**Figure 7.** (a) Transmission characteristics of the 50-period SWS with input/output couplers. (b) $S_{11}$ at the interaction of the wave with an electron beam in the manuscript.

The $S_{11}$ at the interaction of the wave with the electron beam is calculated through the input and reflected power at different frequencies, and the simulation results are shown in Figure 7b, which are less than −15 dB within the frequency 88–95 GHz and are in good agreement with the results in Figure 7a.

## 3. Beam-Wave Interaction Simulation

To evaluate the output performance of the W-band TWT with the proposed dual-beam SV-SL SWS, we used the 3-D particle-in-cell (PIC) CST Particle Studio simulator [21]. In the simulation, we considered two identical 0.15 A electron beams (with a total current of 0.3 A) focused by a uniform magnetic field. And the beam voltage was 17.2 kV. The axis of the electron beam was positioned at 0.075 mm above the SWS. Considering the electron beam tunnel size, the transversal dimension of the beams was 0.8 mm × 0.1 mm (current density of 187.5 A/cm$^2$), corresponding to a broadside filling factor of 40%. The total length of the SWS with 50 periods plus the input/output coupling section was 66 mm. Considering the stable beam transportation without current interception by the beam tunnel walls, the uniform magnetic field should be 0.6 T or higher.

Figure 8a depicts the output power and gain versus input power plot. The simulation maximum output power can reach 200 W with an input power of 0.12 W, and the corresponding gain is 32.2 dB at 90 GHz. By contrast, in Figure 8b, the driving power is set to 0.1 W. The simulation forecasts a maximal gain of 32.9 dB at 90 GHz, and the corresponding output power of 197 W. And the 3 dB bandwidth is 4 GHz.

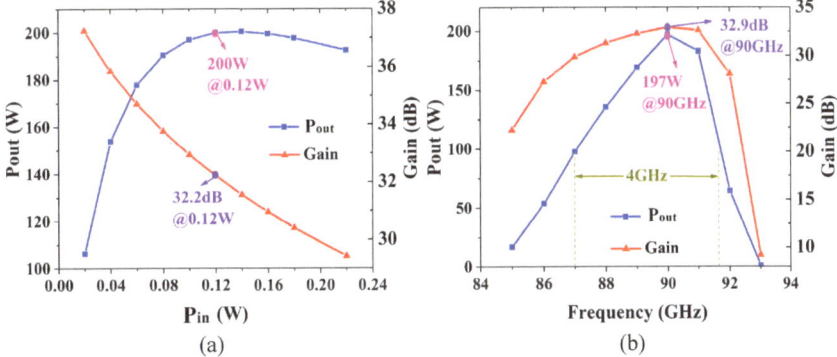

**Figure 8.** (**a**) Output power versus the input power at 90 GHz. (**b**) Output power versus frequency with 0.1 W input power.

As shown in Figure 9a, we simulated the fluctuation of the SWS output signal with time at a 17.2 kV beam voltage, 90 GHz frequency, and 0.1 W input power. The signal is input from port 1, and after wave injection interaction, it is output from port 2, indicating that there is no noticeable oscillation and that it can be amplified stably. The output signal is stable across the whole frequency range, and we observed a clean single-frequency spectrum with no spurious mode excitation, as illustrated in Figure 9b.

Figure 10 shows the corresponding beam phase-space diagram. It indicates the effective beam bunching along the axial direction. Most particles lose energy, and a few electrons gain energy, showing a sufficient beam wave energy transfer.

According to studies [20,22–26], as shown in Table 3, the saturated output power of microstrip-like structures is generally limited by the small size of the structure, which leads to low operating voltages and currents. Moreover, microstrip-like structures are usually held in place by dielectric substrates, which are susceptible to charge accumulation effects, and the meander line SWS using a dielectric substrate needs to take into account the high loss of the intermediate seed layer connecting the dielectric layer to the metal layer in actual processing, so $S_{21}$ in experiments is usually below −30 dB. The SV-SL SWS has a higher operating voltage, and its saturated output power can be more than twice that of the microstrip-like structure at approximately the same 3 dB bandwidth. Since the SV-SL SWS removes the seed and dielectric layers and adopts an all-metal structure, it is easier to process and assemble, and the transmission loss is smaller. Subsequent work should ensure that the operating voltage is reduced as much as possible and the efficiency of the electron interaction is improved with little change in bandwidth.

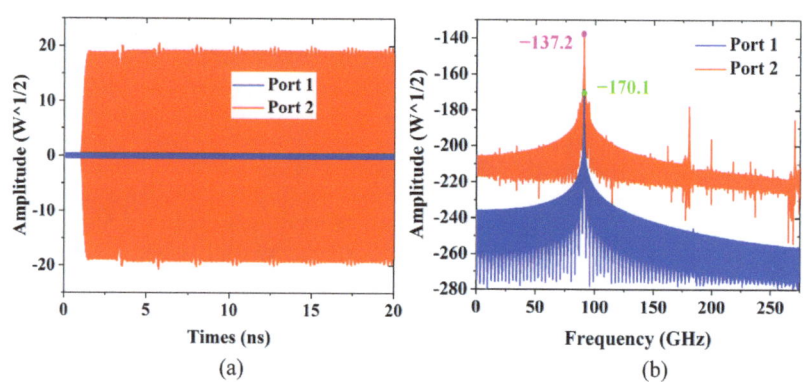

**Figure 9.** (a) Input and output signals versus time for SWS at 17.2 kV beam voltage. (b) The spectrum of the input and output signals.

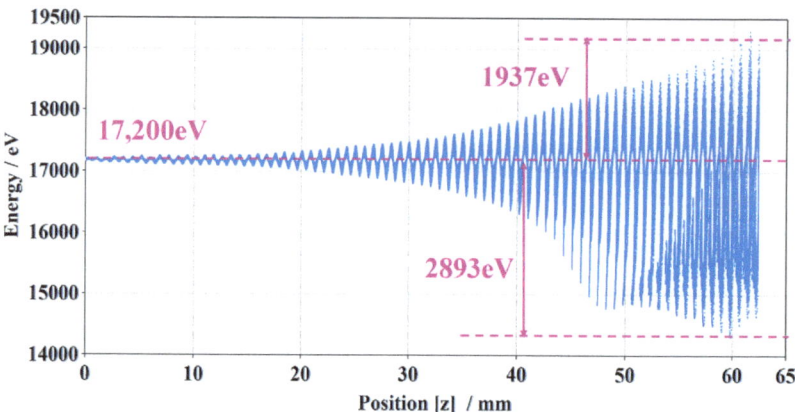

**Figure 10.** Electron beam phase-space diagram.

**Table 3.** Comparison of different planar SWSs in the W-band.

| Type | Operating Parameters | Simulation Output Power, Efficiency and BW |
|---|---|---|
| Metalized ML TWT [20] | 9 kV, 0.028 A | 10 W, 4%, 3 GHz |
| ML SWS with the CVD diamond substrate [22] | 15.6 kV, 0.043 A | 40 W, 5.96%, 5 GHz |
| PML SWS (cylindrical beam) [23] | 6.5 kV, 0.04 A | 36 W, 1.38%, 5 GHz |
| U-shaped ML SWS [24] | 7.1 kV, 0.1 A | 20.77 W, 2.9%, 7 GHz |
| ML SWS with a conformal substrate [25] | 6.55 kV, 0.1 A | 31.4 W, 4.8%, 6 GHz |
| Miniature ML SWS [26] | 14 kV, 0.2 A | 121 W, 4.32%, 3.5 GHz |
| SV-SL SWS | 17.2 kV, 0.3 A | 200 W, 3.63%, 4 GHz |

## 4. Conclusions

In conclusion, we proposed a dual-beam SV-SL SWS and investigated its application in W-band TWT. The SWS was found to have suitable coupling impedance and gentle dispersion characteristics. The simulated transmission characteristics of the 50-pitch structure with the input/output couplers demonstrated good performance in the 87.5–95 GHz frequency range. The transmission loss was substantially lower than the meander-line SWS with dielectric support or substrate. According to the 3-D PIC simulation, when set to a dual beam with 17.2 kV and 0.3 A in total, the maximum output power can reach 200 W at 90 GHz, corresponding to the maximum gain and efficiency of 32.2 dB and 3.63%, respectively. The all-metal SV-SL SWS consists of the following three parts: the metal sheet in the middle is carved by a laser engraving or wire cutting to form a periodic winding groove; the upper and lower metal shells are manufactured by a CNC milling machine or casting; and they are then fixed by resistance welding or brazing. Compared with the microstrip-like structures, SV-SL SWS not only completely solves the charge accumulation problem caused by the dielectric substrate and the loss caused by the seed layer, but it also has good mechanical properties that can account for both thermal properties and large-scale production. As a result, the suggested SV-SL SWS has significant application potential in the W-band TWT.

**Author Contributions:** Conceptualization, Y.W. and S.W.; Methodology, Y.W. and Y.D.; Software, Y.W. and J.G.; Validation, S.W. and D.X.; Formal Analysis, Y.W. and J.G.; Investigation, Y.W.; Resources, S.W., Y.Z., Z.L. and Z.W.; Data Curation, Y.W. and Y.D.; Writing—Original Draft Preparation, Y.W.; Writing—Review and Editing, S.W; Visualization, Y.W.; Supervision, S.W.; Project Administration, S.W.; Funding Acquisition, S.W. All authors have read and agreed to the published version of the manuscript.

**Funding:** This work is supported by the Municipal Government of Quzhou under Grant No. 2022D007, the China Postdoctoral Science Foundation (2022M720664), and the National Natural Science Foundation of China (Grant Nos. 61921002, 61988102, 92163204, 62071087 and 62150052).

**Institutional Review Board Statement:** Not applicable.

**Informed Consent Statement:** Not applicable.

**Data Availability Statement:** No new data were created or analyzed in this study. Data sharing is not applicable to this article.

**Conflicts of Interest:** The authors declare no conflicts of interest.

# References

1. Qiu, J.X.; Levush, B.; Pasour, J.; Katz, A.; Armstrong, C.M.; Whaley, D.R.; Tucek, J.; Kreischer, K.; Gallagher, D. Vacuum tube amplifiers. *IEEE Microw.* **2009**, *10*, 38–51. [CrossRef]
2. Field, M.; Kimura, T.; Atkinson, J.; Gamzina, D.; Luhmann, N.C.; Stockwell, B.; Grant, T.J.; Griffith, Z.; Borwick, R.; Hillman, C.; et al. Development of a 100-W 200-GHz High Bandwidth mm-Wave Amplifier. *IEEE Trans. Electron Devices* **2018**, *65*, 2122–2128. [CrossRef]
3. Ryskin, N.M.; Rozhnev, A.G.; Starodubov, A.V.; Serdobintsev, A.A.; Pavlov, A.M.; Benedik, A.I.; Sinitsyn, N.I. Planar Microstrip Slow-Wave Structure for Low-Voltage V-Band Traveling-Wave Tube with a Sheet Electron Beam. *IEEE Electron Device Lett.* **2018**, *39*, 757–760. [CrossRef]
4. Torgashov, R.A.; Ryskin, N.M.; Rozhnev, A.G.; Starodubov, A.V.; Serdobintsev, A.A.; Pavlov, A.M.; Molchanov, S.Y. Theoretical and Experimental Study of a Compact Planar Slow-Wave Structure on a Dielectric Substrate for the band Traveling-Wave Tube. *Tech. Phys.* **2020**, *65*, 660–665. [CrossRef]
5. Ulisse, G.; Krozer, V.; Ryskin, N.; Starodubov, A.; Serdobintsev, A.; Pavlov, A. Fabrication and measurements of a planar slow wave structure operating in V-band. In Proceedings of the International Vacuum Electronics Conference (IVEC), Busan, Republic of Korea, 28 April–1 May 2019.
6. Yin, P.C.; Xu, J.; Yang, R.C.; Yue, L.N.; Luo, J.J.; Zhang, J.; Wei, Y.Y. An Approach to Focus the Sheet Electron Beam in the Planar Microstrip Line Slow Wave Structure. *IEEE Trans. Electron Devices* **2022**, *69*, 3373–3379. [CrossRef]

7. Ryskin, N.M.; Starodubov, A.V.; Torgashov, R.A.; Rozhnev, A.G.; Pavlov, A.M.; Galushka, V.V.; Serdobintsev, A.A.; Kozhevnikov, I.O.; Ulisse, G.; Krozer, V. Development of a millimeter-band traveling-wave tube with a meander-line microstrip slow wave structure. In Proceedings of the 4th International Conference on Terahertz and Microwave Radiation: Generation, Detection, and Applications (Proc. SPIE), Tomsk, Russia, 24–26 August 2020.
8. Starodubov, A.V.; Serdobintsev, A.A.; Pavlov, A.M.; Galushka, V.V.; Sinev, I.V.; Rozhnev, A.G.; Torgashov, R.A.; Torgashov, G.V.; Ryskin, N.M. Experimental and numerical study of electromagnetic parameters of planar slow-wave structures for millimeter-wave vacuum electronic devices. In Proceedings of the 6th Annual International Symposium on Optics and Biophotonics/22nd Annual Saratov Fall Meeting (SFM)—Laser Physics, Photonic Technologies and Molecular Modeling, Saratov, Russia, 24–29 September 2018.
9. Socuellamos, J.M.; Dionisio, R.; Letizia, R.; Paoloni, C. Experimental Validation of Phase Velocity and Interaction Impedance of Meander-Line Slow-Wave Structures for Space Traveling-Wave Tubes. *IEEE Trans. Microw. Theory Tech.* **2021**, *69*, 2148–2154. [CrossRef]
10. Li, Q.; Yin, P.; Yang, R.; Wei, Y. A Broadband Suspended Coplanar Waveguide Slow-wave Structure for Planar TWTs. In Proceedings of the 2021 22nd International Vacuum Electronics Conference (IVEC), Rotterdam, The Netherlands, 27–30 April 2021.
11. Torgashov, G.V.; Torgashov, R.A.; Titov, V.N.; Rozhnev, A.G.; Ryskin, N.M. Meander-Line Slow-Wave Structure for High-Power Millimeter-Band Traveling-Wave Tubes with Multiple Sheet Electron Beam. *IEEE Electron Device Lett.* **2019**, *40*, 1980–1983. [CrossRef]
12. Yang, R.C. KResearch on High Frequency Systems of Traveling Wave Devices. Ph.D. Thesis, University of Electronic Science and Technology of China, Chengdu, China, 2022.
13. Zhao, C.; Tian, S.; Liu, W.; Liao, X.; Fang, X.; Wang, S. Design and RF Characterization of the Co-Planar Slow Wave Structure for Millimeter-Wave BWO Applications. *IEEE Trans. Electron Devices* **2023**, *71*, 833–839. [CrossRef]
14. Wang, S.; Aditya, S.; Xia, X.; Ali, Z.; Miao, J. On-Wafer Microstrip Meander-Line Slow-Wave Structure at Ka-Band. *IEEE Trans. Electron Devices* **2018**, *65*, 2142–2148. [CrossRef]
15. Ryskin, N.M.; Torgashov, G.V.; Torgashov, R.A.; Ploskih, A.E.; Rozhnev, A.G.; Titov, V.N.; Starodubov, A.V.; Navrotskiy, I.A.; Emelyanov, V.V. Development of Miniature Millimeter-Band Traveling-Wave Tubes with Sheet and Multiple Electron Beams. In Proceedings of the 2020 7th All-Russian Microwave Conference (RMC), Moscow, Russia, 25–27 November 2020.
16. Ruan, C.J.; Zhang, M.W.; Dai, J.; Zhang, C.Q.; Wang, S.Z.; Yang, X.D.; Feng, J.J. W-Band Multiple Beam Staggered Double-Vane Traveling Wave Tube with Broad Band and High Output Power. *IEEE Trans. Plasma Sci.* **2015**, *43*, 2132–2139. [CrossRef]
17. *Introduction of CST Microwave Studio*; Dassault Systemes: Paris, France. Available online: https://www.cst.com/products/cstms (accessed on 1 December 2016).
18. Wang, Y.X.; Wang, S.M.; Dong, Y.; Guo, J.Y.; Xu, D.; Zheng, Y.; Wang, Z.L.; Lu, Z.G.; Gong, H.R.; Duan, Z.Y.; et al. Investigation of a Novel Planar Meander Slot-Line Slow-Wave Structure. *IEEE Electron Device Lett.* **2024**, *45*, 476–479. [CrossRef]
19. Yelizarov, A.A.; Kukharenko, A.S.; Skuridin, A.A. Investigations of a Wideband Metamaterial-based Microstrip Meander Line with Slotted Screen. In Proceedings of the 2019 Thirteenth International Congress on Artificial Materials for Novel Wave Phenomena (Metamaterials), Rome, Italy, 16–19 September 2019.
20. Sengele, S.; Jiang, H.R.; Booske, J.H.; Kory, C.L.; Vander, W.D.; Daniel, W.; Ives, R.L. Microfabrication and Characterization of a Selectively Metallized W-Band Meander-Line TWT Circuit. *IEEE Trans. Electron Devices* **2009**, *56*, 730–737. [CrossRef]
21. *Introduction of CST Particle Studio*; Dassault Systemes: Paris, France. Available online: https://www.cst.com/products/cstps (accessed on 1 December 2016).
22. Galdetskiy, A.; Rakova, E. New slow wave structure for W-band TWT. In Proceedings of the 2017 Eighteenth International Vacuum Electronics Conference (IVEC), London, UK, 24–26 April 2017.
23. Socuéllamos, J.M.; Letizia, R.; Dionisio, R.; Paoloni, C. Pillared Meander Line Slow Wave Structure for W-band Traveling Wave Tubes. In Proceedings of the 2021 22nd International Vacuum Electronics Conference (IVEC), Rotterdam, The Netherlands, 27–30 April 2021.
24. Lu, J.; Yue, L.; Liu, C.; Wang, W.; Zhao, G.; Wei, Y. Design of a W-Band U-shaped Meander-line for Traveling-Wave Tube. In Proceedings of the 2021 22nd International Vacuum Electronics Conference (IVEC), Rotterdam, The Netherlands, 27–30 April 2021.
25. Yue, L.; Shan, W.; Liu, C.; Lu, J.; Wang, W.; Xu, J.; Chen, D.; Zhao, G.; Yin, H.; Guo, G.; et al. A High Interaction Impedance Microstrip Meander-Line with Conformal Dielectric Substrate Layer for a W-Band Traveling-Wave Tube. *IEEE Trans. Electron Devices* **2022**, *69*, 5826–5831. [CrossRef]
26. Torgashov, R.A.; Nozhkin, D.A.; Starodubov, A.V.; Ryskin, N.M. Development and Investigation of a Slow-Wave Structure for a Miniature Multiple-Beam W-Band Traveling Wave Tube. *Commun. Technol. Electron.* **2023**, *68*, 1209–1213. [CrossRef]

**Disclaimer/Publisher's Note:** The statements, opinions and data contained in all publications are solely those of the individual author(s) and contributor(s) and not of MDPI and/or the editor(s). MDPI and/or the editor(s) disclaim responsibility for any injury to people or property resulting from any ideas, methods, instructions or products referred to in the content.

*Communication*

# Resonant Gas Sensing in the Terahertz Spectral Range Using Two-Wire Phase-Shifted Waveguide Bragg Gratings

Yang Cao [1,2,*], Kathirvel Nallappan [2], Guofu Xu [2] and Maksim Skorobogatiy [2,*]

[1] Center for Advanced Laser Technology, Hebei University of Technology, 5340 Xiping Road, Tianjin 300401, China
[2] Engineering Physics, Polytechnique Montréal, C.P. 6079, Succ. Centre-Ville, Montréal, QC H3C 3A7, Canada; kathirvel.nallappan@polymtl.ca (K.N.); guofu.xu@polymtl.ca (G.X.)
* Correspondence: yang.cao@hebut.edu.cn (Y.C.); maksim.skorobogatiy@polymtl.ca (M.S.)

**Abstract:** The development of low-cost sensing devices with high compactness, flexibility, and robustness is of significance for practical applications of optical gas sensing. In this work, we propose a waveguide-based resonant gas sensor operating in the terahertz frequency band. It features micro-encapsulated two-wire plasmonic waveguides and a phase-shifted waveguide Bragg grating (WBG). The modular semi-sealed structure ensures the controllable and efficient interaction between terahertz radiation and gaseous analytes of small quantities. WBG built by superimposing periodical features on one wire shows high reflection and a low transmission coefficient within the grating stopband. Phase-shifted grating is developed by inserting a Fabry–Perot cavity in the form of a straight waveguide section inside the uniform gratings. Its spectral response is optimized for sensing by tailoring the cavity length and the number of grating periods. Gas sensor operating around 140 GHz, featuring a sensitivity of 144 GHz/RIU to the variation in the gas refractive index, with resolution of $7 \times 10^{-5}$ RIU, is developed. In proof-of-concept experiments, gas sensing was demonstrated by monitoring the real-time spectral response of the phase-shifted grating to glycerol vapor flowing through its sealed cavity. We believe that the phase-shifted grating-based terahertz resonant gas sensor can open new opportunities in the monitoring of gaseous analytes.

**Keywords:** terahertz technology; gas sensing; plasmonic waveguide; phase-shifted grating; additive manufacturing

## 1. Introduction

An increasing demand for the monitoring of air quality has promoted the development of high-performance gas sensing devices operating on various chemical and physical principles such as optical, calorimetric, chromatographic, acoustic, as well as electrochemical [1–5]. Among those, optical sensors exhibit unique advantages by being immune to electromagnetic interferences, free of external power supply, capable of operating in harsh environments, and allowing multiplexed remote sensing [6–8]. Furthermore, for various gaseous analytes (e.g., gases, vapors, aerosols), the terahertz band is abundant with spectral fingerprints [9–12], thus opening new opportunities in optical gas sensing. As a complementary technique to the well-established infrared spectroscopy that probes electronic transitions in molecules [13], THz spectroscopy rather probes molecular vibrations, which are particularly pronounced in the gas phase [14]. Additionally, to handle the submillimeter radiation, THz optics are usually much larger than infrared ones, thus enabling novel designs (e.g., integrate with gas cell) and fabrication techniques (e.g., additive manufacturing) of gas sensing devices. However, a significant challenge for gas sensing, particularly at low analyte concentrations, is the weak signal, which prompts the use of long straight gas cells [15,16] or circular multi-pass cells [17,18] to obtain the measurable absorption, thus resulting in large and cumbersome gas sensor systems.

It is, therefore, important to investigate the integrated resonant structures, particularly in the THz band, capable of reducing the size of sensor systems, compared to the free-space systems, without sacrificing sensitivity. One way to achieve this is by using hollow core waveguides filled with gaseous analytes to perform broadband molecular vibration absorption spectroscopy. Such waveguides operate using various guidance principles (e.g., ARROW, bandgap, plasmonic) and offer high field–analyte overlap [19–22] while occupying much smaller volumes (e.g., coiled hollow core fibers [23,24]) than free-space gas cells. They are predominantly used to monitor the frequency-dependent imaginary part (loss) of the analyte Refractive Index (RI). Therefore, for chemical species identification and component differentiation, one usually resorts to the costly THz optical sources supporting stable and broadband operation.

Alternatively, a THz waveguide-based sensor of relatively short length can be designed using various resonant elements in their structures (e.g., Bragg gratings, asymmetric directional couples, integrated Fabry–Perot resonant cavities, and coherent scattering elements [25–31]). Due to the low bandwidth nature of resonant devices, one can then monitor the gaseous analyte RI (mostly its real part) by tracking the spectral position of various singularities using cost-effective THz sources (e.g., resonant tunneling diodes).

Although high sensitivities are readily achievable by both one-dimensional (e.g., photonic crystal cavity on silicon wafer [26]) and two-dimensional resonators (e.g., pillar arrays [29]), it is noted that for most reported optical sensors, the gaseous analyte delivery infrastructure comes as an afterthought. In contrast, in this work, this crucial component is co-engineered with optical ones, thus ensuring the independent efficient operation of both with minimal mutual intrusion for gas sensing. This subtly integrated structure outperforms the conventional open-structured sensors in terms of compactness and performance stability. Particularly, by removing the employment of external gas cells, the proposed sensor is especially suitable for monitoring small quantities of gaseous analytes.

In this work, we propose a real-time resonant THz gas sensor based on phase-shifted waveguide Bragg grating (WBG). At the core of this device is a broadband two-wire plasmonic waveguide formed by metalizing polymer cylinders that are encapsulated within a closed polymer cage. The gaseous analyte flows inside the cage and in the air gap of a two-wire plasmonic waveguide. WBG is formed by a periodic conical pattern imprinted onto one of the cylinders of a two-wire waveguide and is optimized to feature a spectrally broad stopband. Finally, the phase-shifted grating is formed by inserting a Fabry–Perot cavity in the form of a uniform waveguide section in the middle of WBG. The cavity length and the number of grating periods should be chosen to support a single spectrally narrow transmission peak within a broad WBG stopband. The THz spectral response of phase-shifted gratings is then studied for different lengths of a cavity and different refractive indices of gaseous analyte that are filling the semi-sealed cavity. By tracking the position of the transmission peak, our sensor sensitivity near 0.14 THz is found to be ~14.5 GHz/mm for changes in the cavity length, and ~144 GHz/RIU for changes in the analyte RI (real part). A theoretical sensing resolution of ~$7 \times 10^{-5}$ RIU is estimated from the 10 MHz resolution of our spectrometer. Finally, using a continuous-wave (CW) THz spectroscopy system, we experimentally demonstrate the real-time detection of glycerol vapors from an electronic cigarette as an analyte. Namely, when replacing dry air with glycerol vapor in the cavity of a phase-shifted grating module, a shift in the sensor resonant frequency (transmission peak) of ~50 MHz reveals an RI difference of ~$3.5 \times 10^{-4}$ RIU.

Different from the most reported optical gas sensors whose delicate structures are realized using costly infrastructures (e.g., femtosecond laser and deep reactive ion etchers), the proposed gas sensor on a centimeter-scale THz waveguide can be rapidly manufactured using the emerging 3D printing technology with precision and robustness. Owing to the ubiquitous availability of hardware as well as the compact modular design that integrates various crucial elements, we believe that this sensor confronts a lower threshold for entering into production and less challenging engineering problems for operation in practical applications.

## 2. Two-Wire Waveguide Bragg Gratings

Unlike the conventional two-wire metallic waveguides [32,33], the two-wire waveguides used in this work and detailed in [34] feature a modular design with the wires in the form of metalized polymer cylinders encapsulated within a polymer enclosure (see Figure 1a). Such a micro-encapsulated design circumvents the intrinsic engineering defect of conventional one in alignment, and promises mechanical stable, cost-effective, and highly reconfigurable THz optical circuits for various applications (the comparison of transmission spectra is shown in Figure 2d in [34]). The waveguide cross-sectional design, including the wire diameter, the air gap size, as well as the topography of the enclosure, were carefully tailored to ensure the featureless transmission spectra with low insertion loss for a several-centimeter-long waveguide around 140 GHz. Such a design eliminates the presence of spectral ripples and enables distinct measured transmission spectra using THz spectroscopy, thus facilitating the signal identification for gas sensing.

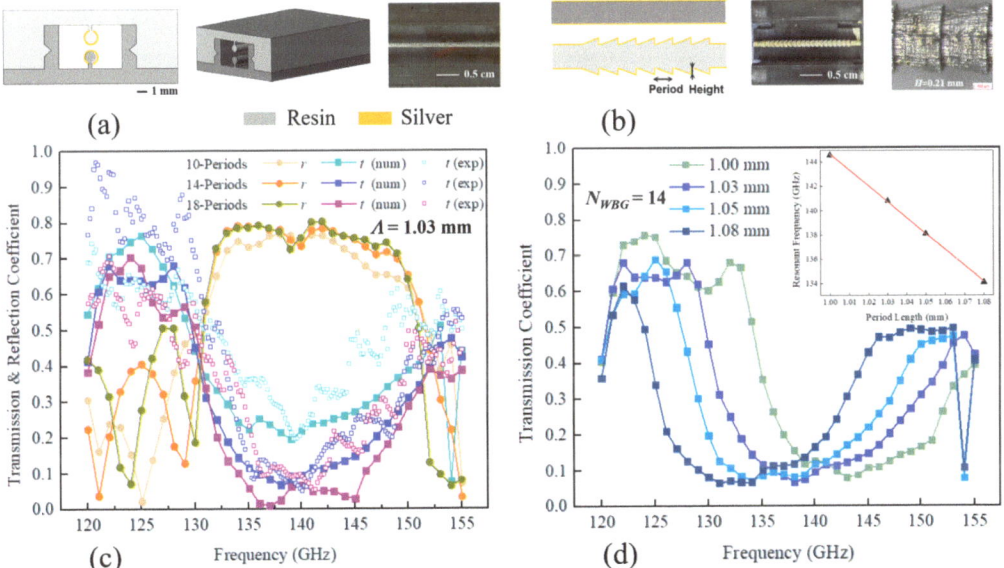

**Figure 1.** Micro-encapsulated two-wire waveguide and WBG fabricated using stereolithography and wet chemistry deposition. (**a**) Schematic of an encapsulated two-wire waveguide. (**b**) The two-wire WBG features a sequence of end-to-end connected truncated cones written on one of the two wires. (**c**) Transmission and reflection spectra of WBGs featuring a different number of periods, $\Lambda = 1.03$ mm. (**d**) Numerical transmission spectra of WBGs for different period lengths, $N_{WBG} = 14$. Inset: The center frequency of a WBG stopband as a function of its period length.

Additionally, the integration of the plasmonic terahertz waveguide and semi-sealed cavity promises the controllable interaction between the supported THz surface plasmon polariton wave and the gaseous analyte flowing through. However, as refractive indices (real part) for most gases are close to one, it is challenging to detect the difference between them, thus necessitating the use of long interaction distances (long gas cells) to accumulate sufficient phase differential between different analytes. In contrast, by using resonant devices like a Fabry–Perot cavity (in this work: realized in the form of a phase-shifted WBG), we can fold the optical path to realize much smaller devices.

Experimentally, we find that the two-wire WBGs featuring a sequence of end-to-end connected truncated cones on one wire was an optimal design that can be printed reliably with high precision and without supports, using a tabletop stereolithography 3D printer (see Appendix A for details in fabrication). In principle, one can further increase the grating strength (stopband bandwidth) by using other geometries such as deep rectangular grooves on both wires. However, it is noted that realizing such designs is challenging due to microstructure deformation induced by the intrinsic cure-through defect of 3D printing and the difficulty of aligning such structures [35].

Specifically, the UV radiation in each exposure not only cures the resin within the top printed layer, but also leaks through the cured layer and solidifies some resin on the other side. Therefore, the resultant cumulative deformation has to be taken into consideration for the grating structure design, as it becomes explicit for prints where geometry changes rapidly from one layer to another. Additionally, the two-wire waveguide components were manually assembled from two complementary 3D-printed parts. When subwavelength features are superimposed on both parts, the postprocessing facet-polishing step can easily lead to their misalignments in practice. Furthermore, the optimal truncated ridge height was found to be ~0.2 mm, enabling a large bandwidth of the stopband, manageable loss in the passband, as well as the reproducible optical performance of printed WBGs (see Figure 1b).

For a stopband center frequency of ~140 GHz, the period of WBGs is found to be $\Lambda = 1.03$ mm. The transmission and reflection spectra for the 2.5 cm long WBGs containing $N_{WBG} = 10, 14, 18$ periods are shown in Figure 1c, with numerical transmission and reflection coefficients in the vicinity of the stopband center frequency reaching <0.1 and >0.75 values, respectively, when the number of periods is over 14. The linear dependence of the stopband center frequency on the grating period $\Lambda$ is shown in Figure 1d for a 14-period structure, with a slope of 131 GHz/mm. Experimentally, the transmission measurements were conducted using a CW-THz spectroscopy system (see Appendix A for details in characterization), and the spectral response of the 3D printed THz WBGs within the grating stopband agrees well with numerical simulation, as seen in Figure 1c. A minimal transmission coefficient of ~0.08 was found for the ~16 GHz wide grating stopband of a 14-period WBG.

Next, we realize a narrow transmission window within the WBG stopband by incorporating a Fabry–Perot cavity, which is a two-wire waveguide section with a length of $L_{F-P} = 2.75$ mm, between two WBG reflectors. The resonance in the Fabry–Perot cavity results in the presence of transmission peaks within the WBG stopband. Experimentally, we find that a 14-period phase-shifted WBG shown in Figure 2a results in a superior performance in terms of the transmission spectra for gas sensing. It is worth noting that the elongation of the grating leads to a narrower transmission peak (~2 GHz bandwidth for a phase-shifted WBG containing 18 periods), but comes at the cost of deteriorated transmission peak intensity (~0.1 transmission coefficient difference between the resonant frequency and other frequencies within the grating stopband), thus posing challenges in identifying the desired transmission peak. Additionally, in a numerical simulation, the bandwidth of the exclusive transmission peak decreases from ~4.7 GHz to ~3.6 GHz when the waveguide length increases from ~0.5$\Lambda$ to ~2.5$\Lambda$. Further reduction in bandwidth by extending the waveguide section is infeasible due to the appearance of multiple spectral singularities within the grating stopband, while the spectral position of the transmission peak with a basically unaffected bandwidth moves toward a lower frequency when the F-P cavity length slightly increases (see Figure 3a).

Because of the standing waves formed inside of the photomixer silicon lenses and free-space cavities of the CW-THz spectroscopy setup, parasite ripples are superimposed on measured transmission spectra [36], posing challenges in identifying the transmission peak of phase-shifted WBGs from the experimental data. To simplify the task, we identify the resonant peak position by subtracting the transmission spectrum of a uniform WBG from the spectra of the phase-shifted WBGs (see Figure 3b). A good correspondence between experiment and theory is found for the spectral position of a transmission peak as a function of the cavity length, with an exception of a small systematic frequency shift of ~2 GHz as seen in Figure 3c. We believe that this consistent discrepancy is mainly attributed to the structural nonuniformity of experimental gratings, which results in the longer equivalent

F-P cavity compared with that of the ideal numerical model. Both in theory and experiment, the dependence is linear, with a slope of ~14.5 GHz/mm.

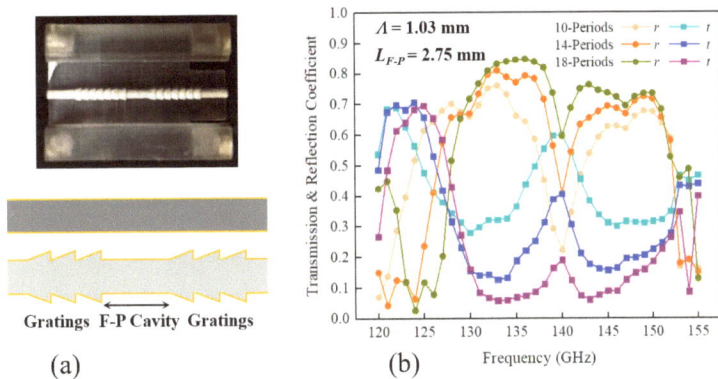

**Figure 2.** Phase-shifted waveguide Bragg grating. (**a**) Schematic and photo of a two-wire waveguide-based phase-shifted WBG. (**b**) Numerical transmission and reflection spectra of a phase-shifted WBG as a function of the number of periods, $\Lambda = 1.03$ mm and $L_{F\text{-}P} = 2.75$ mm.

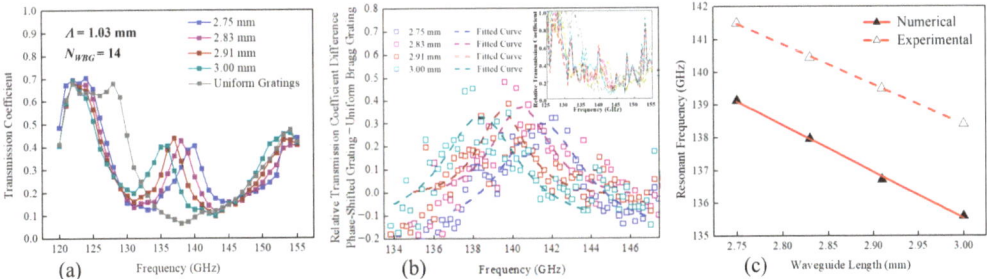

**Figure 3.** Spectral response of phase-shifted WBGs for various cavity lengths with $N_{WBG} = 14$ and $\Lambda = 1.03$ mm. (**a**) Numerical transmission spectra and (**b**) experimental normalized transmission spectra of phase-shifted WBGs. Inset: the transmission spectra of phase-shifted WBGs and a uniform WBG. (**c**) The spectral position of the transmission peak within the WBG stopband as a function of the cavity length.

## 3. Two-Wire Waveguide-Based Resonant Gas Sensor

Finally, we demonstrate real-time THz gas sensing based on our thus-designed phase-shifted WBG. A 2.5 cm long phase-shifted grating module containing a cavity of $L_{F\text{-}P} = 2.75$ mm in the middle of 14-period gratings with $\Lambda = 1.03$ mm was sealed on both ends with polyethylene film ($\alpha < 0.01$ cm$^{-1}$ for a lower-terahertz band) with a thickness of tens of micrometers. In experiments, the addition of such a THz transparent material led to negligible changes in the transmission spectra of this module. To couple with the free-space THz beam for characterization, this module was placed between two 3 cm long featureless two-wire waveguide sections which support broadband operation. The assembled waveguide component was then fitted with conical horn antenna and placed inside the THz spectroscopy setup (see Figure 4). It is noted that three through holes were drilled on the side wall of the enclosure of the phase-shifted grating module for gaseous analyte delivery.

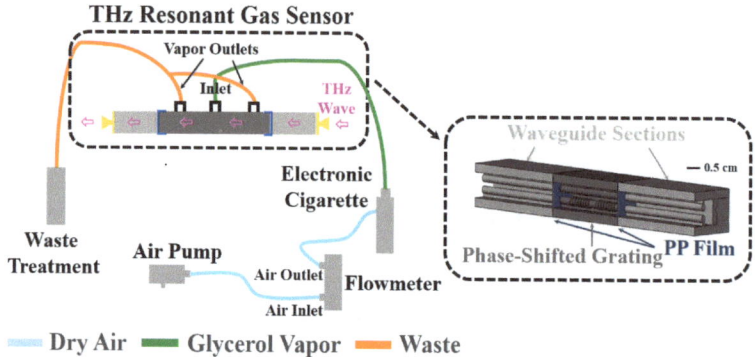

**Figure 4.** Schematic of the experimental setup to fill the cavity hosting two metalized wires with glycerol vapor.

Glycerol is one of the main ingredients of vaping liquid, to which nicotine and flavors are added. The gas mixture generated by electronic cigarettes is notoriously harmful to human health. Specifically, glycerol aerosol alone has been shown to have an impact on the liver and energy metabolism [37]. Therefore, detecting glycerol vapor in air is of practical significance in health management, which was demonstrated by the proposed sensor in this work. In experiments, glycerol vapor generated by an electronic cigarette was introduced into the 0.6 mL volume flow cell through the inlet in the middle of a cell with a constant flow rate of ~20 mL/s, while the waste vapor was removed from the two ends of a flow cell through the outlets for waste treatment. The well-designed location of inlet and outlet openings as well as the short voiding time allow the cavity to completely replace its filled gas in sub seconds, enabling the real-time monitoring of gas RI changes.

The numerical simulations of the independent phase-shifted grating module predict that the spectral position of a transmission peak is linear with the gaseous analyte RI with the corresponding sensitivity of 144 GHz/RIU (see Figure 5a). Given the 10 MHz resolution of our CW-THz spectrometer, the theoretical resolution of our sensor is then estimated to be $7 \times 10^{-5}$ RIU, which is as much as an order of magnitude lower than the RI difference between most common gases (e.g., the difference is on the level of $10^{-3}$ to $10^{-4}$ RIU) [38]. In experiments, the transmission spectra of a phase-shifted WBG with an empty cavity, the cavity with dry airflow, and the cavity with glycerol vapor flow were measured subsequently. In dynamic measurements covering the spectral range of a transmission peak, the scanning time for a single data point was ~10 s to alleviate the impact of the inherent latency of a CW spectroscopy system using lock-in acquisition, and to ensure fine spectra with 10 MHz resolution. The center position of a transmission peak was found by first fitting a data cloud of the normalized phase-shifted WBG transmission spectra within the grating stopband using smooth Lorentzian lineshapes,

$$T_{\text{Norm}}(v, v_{center}, \Delta v, A, T_0) = T_0 + A \frac{\Delta v}{4(v - v_{center})^2 + \Delta v^2} \quad (1)$$

and then finding the spectral position of the fit maximum $v_{center}$, similarly to what is shown in Figure 3b.

A typical sensor readout is presented in Figure 5b from which we see that the spectral position of the transmission peak is relatively stable in continuously recorded transmission spectra of the same analyte. Additionally, for an empty cell or a cell with a flow of dry air, the position of the transmission maximum also remains practically unchanged, indicating the immunity of the proposed sensor to changes in gas flow rate. At the same time, when introducing the glycerol vapor, the transmission peak shifts by ~50 MHz, which corresponds to the RI change of ~$3.5 \times 10^{-4}$ compared to that of dry air. Highly consistent experimental results were obtained in each measurement of this sensor. Owing to its compact integrated structure and insensitivity to the environment change, this two-wire waveguide-based sensor can find its practical applications in gas sensing by simply replacing the external infrastructure for gas delivery (see the setup out of the black dotted region in Figure 4). For instance, one can detect the concentration of explosive or toxic gas flowing in pipelines or dispersed in the air remotely in petrochemical industry.

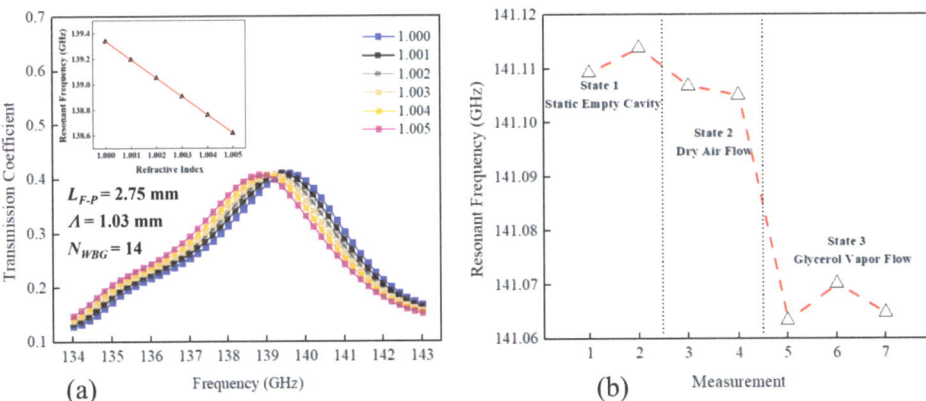

Figure 5. The spectral response of a phase-shifted WBG with gaseous analytes of different RIs in the cavity. (a) Numerical transmission spectrum of a phase-shifted WBG in the vicinity of a resonant peak for different values of gaseous analyte RI. Inset: The spectral position of the transmission peak as a function of the analyte RI. A slope of ~144 GHz/RIU can be found in the linear fit (red line). (b) Experimental time dependence of the spectral position of the transmission peak. Its variation can be found in the red dotted line.

## 4. Discussion

In this work, we propose micro-encapsulated two-wire plasmonic waveguide-based phase-shifted Bragg gratings and demonstrate their applications in real-time THz gas sensing. End-to-end connected truncated cones with a ridge height of ~0.2 mm superposed on one of the two wires were chosen as an optimal WBG design. Low transmission and high reflection coefficients were found within the ~16 GHz wide stopband of such WBGs. Phase-shifted WBG featuring a Fabry–Perot cavity was then developed by placing a uniform waveguide section in the center of a WBG. A single narrow transmission peak of ~3.6 GHz (HFWM) bandwidth in the middle of a WBG stopband was realized by using a ~2.75 mm long cavity flanked on both sides by two seven-period WBGs with a Q-factor of ~39. The theoretical sensitivity of the peak spectral position to changes in the RI of gaseous analytes inside the 2.5 cm long phase-shifted WBG is estimated to be 144 GHz/RIU. The response of our sensor to glycerol vapor flow at low concentrations was then verified in a proof-of-concept time-resolved experiment, which reliably detected the displacement of dry air by glycerol vapor with a resultant RI change of ~$3.5 \times 10^{-4}$ RIU.

For future work, we note that higher sensitivity sensor designs are readily achievable by moving the sensor operational frequency to higher frequencies [39], while also increasing the number of periods in the WBG to reduce the spectral width of a transmission peak. The long-term stability of the proposed sensor also needs to be characterized and further optimized for practical applications. Additionally, considering the modular and reconfigurable design of micro-encapsulated two-wire waveguide components, the sensing of selectivity is readily available via collaboration with THz waveguide-based spectroscopy [19] for various monitoring applications of gaseous analytes such as trace gas analysis and detection [40].

**Author Contributions:** Conceptualization, Y.C. and M.S.; methodology, Y.C.; software, Y.C. and K.N.; validation, Y.C., K.N. and G.X.; formal analysis, Y.C.; investigation, Y.C.; resources, M.S.; data curation, Y.C.; writing—original draft preparation, Y.C.; writing—review and editing, M.S.; visualization, Y.C.; supervision, M.S.; project administration, M.S.; funding acquisition, Y.C. and M.S. All authors have read and agreed to the published version of the manuscript.

**Funding:** This research was funded by Canada Research Chair in Ubiquitous Terahertz Photonics, grant number 34633 and National Natural Science Foundation of China, grant number 62205100.

**Institutional Review Board Statement:** Not applicable.

**Informed Consent Statement:** Not applicable.

**Data Availability Statement:** Data associated with this research are available and can be obtained from authors upon reasonable request.

**Acknowledgments:** We thank our technicians Jean-Paul Lévesque and Yves Leblanc for their assistance.

**Conflicts of Interest:** The authors declare no conflict of interest.

## Appendix A

The proposed micro-encapsulated two-wire plasmonic waveguide components were split into two complementary parts, which exhibit one metalized wire attached to a half dielectric cage. We fabricated the dielectric support of such designed structures using an SLA 3D printer (Asiga Freeform PRO2) with the waveguide direction corresponding to its Z-axis. To suppress deformations associated with the cure-through-resin effect, the parts having periodical subwavelength features superimposed on wires were printed along the single direction from the cone smaller base toward its larger base using the finest layer thickness (10 μm). After 3D printing, we protected the cage's inner surface and then deposited a silver layer onto the uncovered wire support through wet chemistry deposition. Finally, two selectively metalized parts were assembled into a two-wire waveguide component by matching the V-shaped groove and ridge on each cage.

The proposed THz two-wire waveguide-based components, including uniform and phase-shifted WBGs, were characterized using a free-space CW-THz spectroscopy system (Toptica Photonics TeraScan 1550). Tunable THz radiation corresponding to the beat frequency between two C-band distributed feedback lasers was generated in the emitter photomixer. Transmitted through waveguide components, the amplitude of the THz signal was recorded by the receiver photomixer using lock-in detection. In experiments, WBG modules were placed between two two-wire waveguide sections fitted with WR6.5 conical horn antennas (Virginia Diodes). Their transmission coefficients were calculated by comparing the measured transmission spectra of this assembly with the reference corresponding to a two-wire waveguide assembly of the same length.

## References

1. Liu, X.; Cheng, S.; Liu, H.; Hu, S.; Zhang, D.; Ning, H. A survey on gas sensing technology. *Sensors* **2012**, *12*, 9635–9665. [CrossRef] [PubMed]
2. Korotcenkov, G.; Han, S.D.; Stetter, J.R. Review of electrochemical hydrogen sensors. *Chem. Rev.* **2009**, *109*, 1402–1433. [CrossRef] [PubMed]
3. Pathak, A.K.; Swargiary, K.; Kongsawang, N.; Jitpratak, P.; Ajchareeyasoontorn, N.; Udomkittivorakul, J.; Viphavakit, C. Recent Advances in Sensing Materials Targeting Clinical Volatile Organic Compound (VOC) Biomarkers: A Review. *Biosensors* **2023**, *13*, 114. [CrossRef] [PubMed]
4. Chen, C.; Jiang, M.; Luo, X.; Tai, H.; Jiang, Y.; Yang, M.; Xie, G.; Su, Y. Ni-Co-P hollow nanobricks enabled humidity sensor for respiratory analysis and human-machine interfacing. *Sens. Actuators B Chem.* **2022**, *370*, 132441. [CrossRef]
5. Ouyang, Y.; Li, X.; Li, S.; Peng, P.; Yang, F.; Wang, Z.L.; Wei, D. Opto-iontronic coupling in triboelectric nanogenerator. *Nano Energy* **2023**, *116*, 108796. [CrossRef]
6. Hodgkinson, J.; Tatam, R.P. Optical gas sensing: A review. *Meas. Sci. Technol.* **2012**, *24*, 012004. [CrossRef]
7. Li, Z.; Yan, T.; Fang, X. Low-dimensional wide-bandgap semiconductors for UV photodetectors. *Nat. Rev. Mater.* **2023**, *8*, 587–603. [CrossRef]

8. Li, Z.; Liu, X.; Zuo, C.; Yang, W.; Fang, X. Supersaturation-controlled growth of monolithically integrated lead-free halide perovskite single-crystalline thin film for high-sensitivity photodetectors. *Adv. Mater.* **2021**, *33*, 2103010. [CrossRef]
9. Mittleman, D.M.; Jacobsen, R.H.; Neelamani, R.; Baraniuk, R.G.; Nuss, M.C. Gas sensing using terahertz time-domain spectroscopy. *Appl. Phys. B Lasers Opt.* **1998**, *67*, 379–390. [CrossRef]
10. Yang, L.; Guo, T.; Zhang, X.; Cao, S.; Ding, X. Toxic chemical compound detection by terahertz spectroscopy: A review. *Rev. Anal. Chem.* **2018**, *37*, 20170021. [CrossRef]
11. Neumaier, P.X.; Schmalz, K.; Borngräber, J.; Wylde, R.; Hübers, H.W. Terahertz gas-phase spectroscopy: Chemometrics for security and medical applications. *Analyst* **2015**, *140*, 213–222. [CrossRef]
12. Rothbart, N.; Holz, O.; Koczulla, R.; Schmalz, K.; Hübers, H.W. Analysis of human breath by millimeter-wave/terahertz spectroscopy. *Sensors* **2019**, *19*, 2719. [CrossRef] [PubMed]
13. Du, Z.; Zhang, S.; Li, J.; Gao, N.; Tong, K. Mid-infrared tunable laser-based broadband fingerprint absorption spectroscopy for trace gas sensing: A review. *Appl. Sci.* **2019**, *9*, 338. [CrossRef]
14. McIntosh, A.I.; Yang, B.; Goldup, S.M.; Watkinson, M.; Donnan, R.S. Terahertz spectroscopy: A powerful new tool for the chemical sciences? *Chem. Soc. Rev.* **2012**, *41*, 2072–2082. [CrossRef] [PubMed]
15. Bigourd, D.; Cuisset, A.; Hindle, F.; Matton, S.; Fertein, E.; Bocquet, R.; Mouret, G. Detection and quantification of multiple molecular species in mainstream cigarette smoke by continuous-wave terahertz spectroscopy. *Opt. Lett.* **2006**, *31*, 2356–2358. [CrossRef] [PubMed]
16. Vaks, V.L.; Anfertev, V.A.; Balakirev, V.Y.; Basov, S.A.; Domracheva, E.G.; Illyuk, V.; Kupriyanov, P.V.; Pripolzin, S.I.; Chernyaeva, M.B. High resolution terahertz spectroscopy for analytical applications. *Phys.-Uspekhi* **2020**, *63*, 708. [CrossRef]
17. Podobedov, V.B.; Plusquellic, D.F.; Fraser, G.T. Investigation of the water-vapor continuum in the THz region using a multipass cell. *J. Quant. Spectrosc. Radiat. Transf.* **2005**, *91*, 287–295. [CrossRef]
18. Rothbart, N.; Schmalz, K.; Hübers, H.W. A compact circular multipass cell for millimeter-wave/terahertz gas spectroscopy. *IEEE Trans. Terahertz Sci. Technol.* **2019**, *10*, 9–14. [CrossRef]
19. Laman, N.; Harsha, S.S.; Grischkowsky, D.; Melinger, J.S. High-resolution waveguide THz spectroscopy of biological molecules. *Biophys. J.* **2008**, *94*, 1010–1020. [CrossRef]
20. Theiner, D.; Limbacher, B.; Jaidl, M.; Ertl, M.; Hlavatsch, M.; Unterrainer, K.; Mizaikoff, B.; Darmo, J. Flexible terahertz gas sensing platform based on substrate-integrated hollow waveguides and an opto-electronic light source. *Opt. Express* **2023**, *31*, 15983–15993. [CrossRef]
21. Lu, J.Y.; You, B.; Wang, J.Y.; Jhuo, S.S.; Hung, T.Y.; Yu, C.P. Volatile gas sensing through terahertz pipe waveguide. *Sensors* **2020**, *20*, 6268. [CrossRef] [PubMed]
22. Kurt, H.; Citrin, D.S. Photonic crystals for biochemical sensing in the terahertz region. *Appl. Phys. Lett.* **2005**, *87*, 041108. [CrossRef]
23. Tutuncu, E.; Kokoric, V.; Wilk, A.; Seichter, F.; Schmid, M.; Hunt, W.E.; Manuel, A.M.; Mirkarimi, P.; Alameda, J.B.; Carter, J.C.; et al. Fiber-coupled substrate-integrated hollow waveguides: An innovative approach to mid-infrared remote gas sensors. *ACS Sens.* **2017**, *2*, 1287–1293. [CrossRef] [PubMed]
24. Parry, J.P.; Griffiths, B.C.; Gayraud, N.; McNaghten, E.D.; Parkes, A.M.; MacPherson, W.N.; Hand, D.P. Towards practical gas sensing with micro-structured fibres. *Meas. Sci. Technol.* **2009**, *20*, 075301. [CrossRef]
25. Shi, X.; Zhao, Z.; Han, Z. Highly sensitive and selective gas sensing using the defect mode of a compact terahertz photonic crystal cavity. *Sens. Actuators B Chem.* **2018**, *274*, 188–193. [CrossRef]
26. Chen, T.; Han, Z.; Liu, J.; Hong, Z. Terahertz gas sensing based on a simple one-dimensional photonic crystal cavity with high-quality factors. *Appl. Opt.* **2014**, *53*, 3454–3458. [CrossRef]
27. Qin, J.; Zhu, B.; Du, Y.; Han, Z. Terahertz detection of toxic gas using a photonic crystal fiber. *Opt. Fiber Technol.* **2019**, *52*, 101990. [CrossRef]
28. You, B.; Lu, J.Y.; Yu, C.P.; Liu, T.A.; Peng, J.L. Terahertz refractive index sensors using dielectric pipe waveguides. *Opt. Express* **2012**, *20*, 5858–5866. [CrossRef]
29. You, B.; Takaki, R.; Hsieh, C.C.; Iwasa, R.; Lu, J.Y.; Hattori, T. Terahertz Bragg resonator based on a mechanical assembly of metal grating and metal waveguide. *J. Light. Technol.* **2020**, *38*, 3701–3709. [CrossRef]
30. Liu, Y.; Feng, J.; Li, Z.; Luo, J.; Kou, T.; Yuan, S.; Li, M.; Liu, Y.; Peng, Y.; Wang, S.; et al. Double-groove terahertz chirped grating waveguide tube for gas pressure detection. *Laser Phys. Lett.* **2019**, *16*, 056202. [CrossRef]
31. Campanella, C.E.; Cuccovillo, A.; Campanella, C.; Yurt, A.; Passaro, V.M. Fibre Bragg grating based strain sensors: Review of technology and applications. *Sensors* **2018**, *18*, 3115. [CrossRef] [PubMed]
32. Wang, K.; Mittleman, D.M. Metal wires for terahertz wave guiding. *Nature* **2004**, *432*, 376–379. [CrossRef]
33. Mbonye, M.; Mendis, R.; Mittleman, D.M. A terahertz two-wire waveguide with low bending loss. *Appl. Phys. Lett.* **2009**, *95*, 233506. [CrossRef]
34. Cao, Y.; Nallappan, K.; Guerboukha, H.; Xu, G.; Skorobogatiy, M. Additive manufacturing of highly reconfigurable plasmonic circuits for terahertz communications. *Optica* **2020**, *7*, 1112–1125. [CrossRef]
35. Cao, Y.; Nallappan, K.; Xu, G.; Skorobogatiy, M. Add drop multiplexers for terahertz communications using two-wire waveguide-based plasmonic circuits. *Nat. Commun.* **2022**, *13*, 4090. [CrossRef]

36. Cao, Y.; Nallappan, K.; Guerboukha, H.; Gervais, T.; Skorobogatiy, M. Additive manufacturing of resonant fluidic sensors based on photonic bandgap waveguides for terahertz applications. *Opt. Express* **2019**, *27*, 27663–27681. [CrossRef] [PubMed]
37. Ali, N.; Xavier, J.; Engur, M.; Mohanan, P.V.; Serna, J.B. The impact of e-cigarette exposure on different organ systems: A review of recent evidence and future perspectives. *J. Hazard. Mater.* **2023**, *457*, 131828. [CrossRef]
38. Sang, B.H.; Jeon, T.I. Pressure-dependent refractive indices of gases by THz time-domain spectroscopy. *Opt. Express* **2016**, *24*, 29040–29047. [CrossRef]
39. Poulin, M.; Giannacopoulos, S.; Skorobogatiy, M. Surface wave enhanced sensing in the terahertz spectral range: Modalities, materials, and perspectives. *Sensors* **2019**, *19*, 550. [CrossRef]
40. Consolino, L.; Bartalini, S.; Beere, H.E.; Ritchie, D.A.; Vitiello, M.S.; Natale, P.D. THz QCL-based cryogen-free spectrometer for in situ trace gas sensing. *Sensors* **2013**, *13*, 3331–3340. [CrossRef]

**Disclaimer/Publisher's Note:** The statements, opinions and data contained in all publications are solely those of the individual author(s) and contributor(s) and not of MDPI and/or the editor(s). MDPI and/or the editor(s) disclaim responsibility for any injury to people or property resulting from any ideas, methods, instructions or products referred to in the content.

*Communication*

# A Novel Staggered Double-Segmented Grating Slow-Wave Structure for 340 GHz Traveling-Wave Tube

Zechuan Wang [1,†], Junwan Zhu [1,†], Zhigang Lu [1,2,*], Jingrui Duan [1], Haifeng Chen [1], Shaomeng Wang [1], Zhanliang Wang [1], Huarong Gong [1] and Yubin Gong [1]

[1] National Key Laboratory of Science and Technology on Vacuum Electronics, School of Electronic Science and Engineering, University of Electronic Science and Technology of China, No. 2006 Xiyuan Avenue, High-Tech District (West District), Chengdu 611731, China; 202122022214@std.uestc.edu.cn (Z.W.); zhujunwan@yeah.net (J.Z.)

[2] Yangtze Delta Region Institute (Huzhou), University of Electronic Science and Technology of China, Huzhou 313001, China

\* Correspondence: lzhgchnn@uestc.edu.cn

† These authors contributed equally to this work.

**Abstract:** In this paper, a novel staggered double-segmented grating slow-wave structure (SDSG-SWS) is developed for wide-band high-power submillimeter wave traveling-wave tubes (TWTs). The SDSG-SWS can be considered as a combination of the sine waveguide (SW) SWS and the staggered double-grating (SDG) SWS; that is, it is obtained by introducing the rectangular geometric ridges of the SDG-SWS into the SW-SWS. Thus, the SDSG-SWS has the advantages of the wide operating band, high interaction impedance, low ohmic loss, low reflection, and ease of fabrication. The analysis for high-frequency characteristics shows that, compared with the SW-SWS, the SDSG-SWS has higher interaction impedance when their dispersions are at the same level, while the ohmic loss for the two SWSs remains basically unchanged. Furthermore, the calculation results of beam–wave interaction show that the output power is above 16.4 W for the TWT using the SDSG-SWS in the range of 316 GHz–405 GHz with a maximum power of 32.8 W occurring at 340 GHz, whose corresponding maximum electron efficiency is 2.84%, when the operating voltage is 19.2 kV and the current is 60 mA.

**Keywords:** traveling-wave tube; slow-wave structure; staggered double-segmented grating; high interaction impedance; low ohmic loss

## 1. Introduction

As a popular research topic in the field of electromagnetic wave science, benefiting from its superiority in permeability, controllability, and transmissibility, the terahertz wave is widely utilized in communication systems, imaging fields, and biomedical fields [1]. In the field of terahertz science, how to generate the terahertz wave is a very key issue [2–4]. As highly effective broadband high-power signal sources, vacuum electronic devices (VEDs) are widely used. As one of many VEDs, the traveling-wave tube (TWT) is widely used as a broadband high-power amplifier. As the main site of interaction between an electromagnetic wave and an electron beam, the slow-wave structure (SWS) has a large impact on the performance of the TWT [5,6].

Currently, the main SWSs used for TWTs at 340 GHz include the folded waveguide (FW) [7–9], staggered double-grating (SDG) [10–12], sine waveguide (SW) [13–15], and deformations of the above three SWSs. However, with the reduction in SWS size caused by the increase in operating frequency, the sizes of electron beam tunnels become progressively smaller, which will limit the improvement of beam current and output power; at the same time, the ohmic loss of the metal also increases, due to skin depth and fabrication accuracy. In order to reduce the impact of these two problems, on the one hand, a sheet electron beam [16–18] that has a larger dimension should be considered for interaction with the

SWS to reduce the impact of the reduced size of the SWS, and, on the other hand, a new SWS with low transmission loss should be chosen as the site of beam–wave interaction.

Due to the natural sheet beam tunnel, low ohmic loss, and wide operating band, the SW has become a research hotspot in recent years [13,15]. Thus, SW is a suitable SWS for submillimeter wave TWTs [19]. However, the relatively low interaction impedance of the SW-SWS will affect the improvement of SW-TWT performance in terms of output power, gain, and electron efficiency. Therefore, improving the interaction impedance of the SW-SWS while retaining the advantages of a wide operating band and low ohmic loss is a worthwhile research issue [14,19,20].

The SW-SWS, as a modification of the SDG-SWS, has the characteristics of low reflection and low ohmic loss, compared with the SDG-SWS, but its interaction impedance is lower than that of the SDG-SWS. Therefore, based on the comprehensive analysis of the SW-SWS, with low ohmic loss, and the SDG-SWS, with high interaction impedance, a novel staggered double-segmented grating (SDSG) SWS is proposed by innovatively introducing the rectangular geometric ridges of the SDG-SWS into the SW-SWS. The new SDSG-SWS combines the advantages of the above two SWSs and maintains the characteristics of lower ohmic loss and higher interaction impedance.

The remainder of the article is arranged as follows: The high-frequency electromagnetic characteristics of the SDSG-SWS are analyzed in Section 2; Section 3 describes the particle-in-cell (PIC) simulation of the beam–wave interaction of TWT using the SDGS-SWS and analyzes the results; in Section 4, the related conclusions are drawn.

## 2. Design and Analysis

For the SWSs, the capacity of the electromagnetic wave to exchange energy with the electron beam is generally characterized by the interaction impedance $K_c$, which is defined as:

$$K_c = \frac{E_{zn} E_{zn}^*}{2\beta_n^2 v_g U} \tag{1}$$

Here, $E_{zn}$ is the longitudinal electric field component of the $n$th spatial harmonic, $E_{zn}^*$ is the conjugate value for $E_{zn}$, $U$ is the system energy storage per unit length, $v_g$ is the group velocity of electromagnetic wave transmission, and the propagation constant for the $n$th spatial harmonic is $\beta_n$.

For the TWTs, obtaining a greater output power over a wide operating frequency band is required under the same operating conditions. Therefore, the dispersion of the SWS is designed at the same level for the comparison of TWTs with different SWSs; that is, the transmission characteristics for the different SWSs are the same. Of course, the synchronous voltage, also known as the operating voltage, is the same. Subsequently, the research on the electric field distribution of SWSs is a key focus, which is closely related to interaction impedance and ohmic loss. According to Formula (1), and based on the previous analysis, for the different SWSs, the denominator of (1), which is dominated by the dispersion, is almost the same, and improving the interaction impedance depends entirely on the electric field distribution of SWSs.

Figure 1 shows the distribution of the longitudinal electric field $E_Z$ for the SW-SWS and SDG-SWS, respectively. As observed in Figure 1a, the $E_Z$ is mainly concentrated at the bend of the SW-SWS, which is comparable to that of the SDG-SWS in Figure 1b. However, the $E_Z$ in region I of the SDG-SWS is significantly stronger than that in the same region of the SW-SWS, which is due to the SDG-SWS having rectangular geometric ridges. Thus, a new idea was proposed: to introduce the rectangular geometric ridges of the SDG-SWS into the SW-SWS in order to improve the interaction impedance of the SW-SWS. Based on the above idea, the SDSG-SWS is proposed.

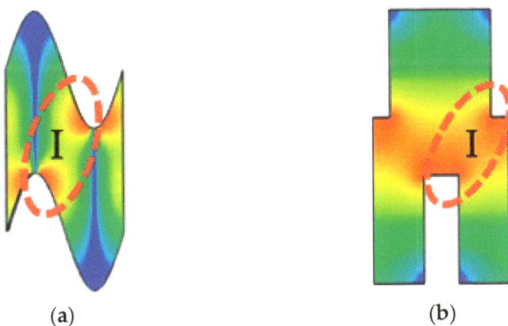

**Figure 1.** Distribution of longitudinal electric field $E_Z$ for (**a**) SW-SWS and (**b**) SDSG-SWS.

Figure 2a,b present the three-dimensional solid models (removing the side wall at X-max) of the SDSG-SWS and SW-SWS, respectively, while Figure 2c,d show the left view and sectional view, respectively, in the y-o-z plane of the SDSG-SWS with the sheet electron beam (the red part is the sheet electron beam). As observed in Figure 2, both the SDSG-SWS and SW-SWS have equal cross-sectional features ($b \times w$), and the SDSG-SWS has the same rectangular geometric ridges as the SDG-SWS at the top and bottom of the metal grating. To better illustrate the origin of the SDSG-SWS, the detailed evolution from SW-SWS to SDSG-SWS is presented in Figure 3.

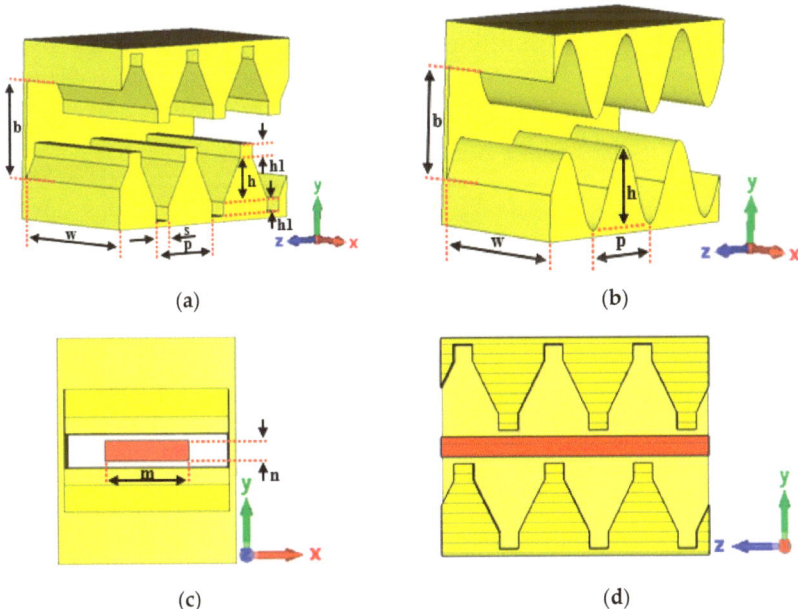

**Figure 2.** Solid models without side wall at X-max of (**a**) SDSG-SWS and (**b**) SW-SWS; (**c**) left-view and (**d**) sectional view in the y-o-z plane of SDSG-SWS with the sheet beam.

The transition structure presented in Figure 3b with the same equal cross-section characteristics as the SW-SWS can be obtained by introducing the rectangular geometric ridges of the SDG-SWS at the top and bottom of the sinusoidal-shaped grating of the SW-SWS presented in Figure 3a. Considering the manufacturing accuracy and processing difficulty of the transition structure in the submillimeter wave band, the sinusoidal profile of the grating in Figure 3b is replaced by a linear profile, and the SDSG-SWS presented in Figure 3c is obtained. The SDSG-SWS is the combination of the SW-SWS and SDG-SWS.

The SDSG-SWS is obtained by introducing the rectangular geometric ridges of the SDG-SWS while keeping the cross-sectional characteristics of the SW-SWS. Therefore, it can be predicted that the SDSG-SWS should have almost the same ohmic loss and dispersion characteristics as the SW-SWS, but its interaction impedance should be higher than that of the SW-SWS.

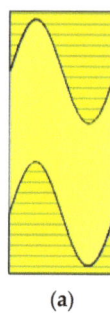
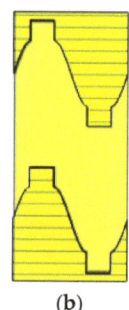
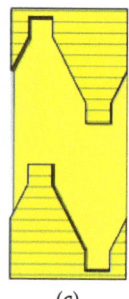

(a) (b) (c)

**Figure 3.** Sectional views in the *y-o-z* plane of (**a**) SW-SWS, (**b**) transition structure, and (**c**) SDSG-SWS.

In order to verify the above speculation and illustrate the advantages of the SDSG-SWS over the SW-SWS, the normalized phase velocities of both SWSs should be kept at the same level within the same frequency band, which is a prerequisite for the comparison. The optimized parameters are presented in Table 1. The electromagnetic characteristics of the two SWSs are calculated using the 3D simulation software Ansoft High Frequency Structure Simulator. The dispersion, attenuation constant, and interaction impedance calculated are presented in Figures 4–6.

**Table 1.** Optimal parameters of SDSG-SWS and SW-SWS.

| Parameter | Value (mm) | |
|---|---|---|
| | SDSG-SWS | SW-SWS |
| $p$ | 0.282 | 0.282 |
| $b$ | 0.36 | 0.37 |
| $w$ | 0.49 | 0.49 |
| $h$ | 0.16 | 0.27 |
| $h1$ | 0.05 | / |
| $s$ | 0.06 | / |
| $m$ | 0.25 | 0.25 |
| $n$ | 0.1 | 0.1 |

Figure 4 shows the dispersion curves for the SDSG-SWS and SW-SWS. The results show that, in a fairly wide frequency range, the normalized phase velocities are essentially the same when their parameters are optimized. Based on these, Figure 5 shows the attenuation constants of the SDSG-SWS and SW-SWS, and the results show that the attenuation constants of both SWSs are also essentially equal for the same dispersion. The ohmic loss of SWSs is represented by the attenuation constant. The results prove that the introduction of rectangular geometric ridges does not change the low ohmic loss characteristics of the SW-SWS under the same dispersion.

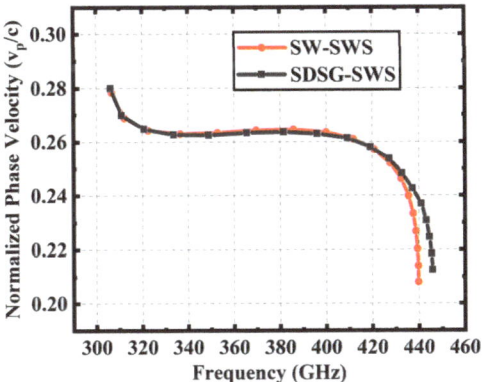

**Figure 4.** Dispersion curves for SDSG-SWS and SW-SWS.

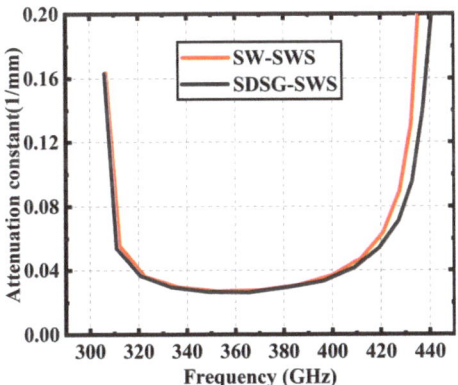

**Figure 5.** Attenuation constant curves of SDSG-SWS and SW-SWS.

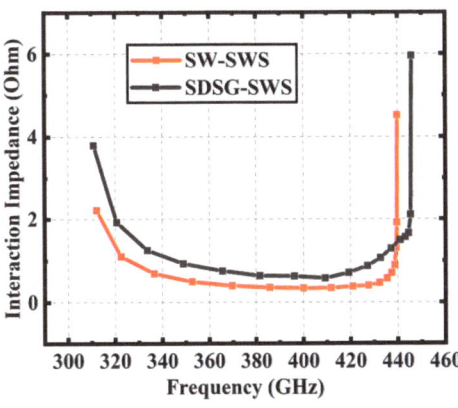

**Figure 6.** Average interaction impedance curves of SDSG-SWS and SW-SWS.

Comparison of interaction impedance for both SWSs is presented in Figure 6. The results show that the SDSG-SWS has a higher interaction impedance compared with the SW-SWS. In the band range of 310 GHz–420 GHz, the minimum value of interaction impedance is 0.57 Ohm for the SDSG-SWS and 0.34 Ohm for the SW-SWS, an improvement of 59.6%. The improved interaction impedance means that the electric field can better exchange energy with the electron beam, which can effectively improve the power, gain, and electron efficiency of the TWT.

To further illustrate that the enhancement of the $E_Z$ is the reason for the enhancement of the SDSG-SWS's interaction impedance, by using the CST eigenmode solver, the $E_Z$ along the black lines shown in Figure 7 (A-B and C-D) is calculated, and the results are shown in Figure 8. It can be observed that, for the SW-SWS, the $E_Z$ is strongest near the bend and gradually decreases as the distance from the bend increases; the trend of the $E_Z$ for the SDSG-SWS is comparable to that for the SW-SWS. However, the calculated value of the $E_Z$ is higher than that of the SW-SWS from Figure 8. The results indicate that the SDSG-SWS has a larger longitudinal electric field $E_Z$ at the place of its electron beam tunnel.

**Figure 7.** Distribution of $E_Z$ in the y-o-z plane of (**a**) SDSG-SWS, (**b**) SW-SWS at 340 GHz.

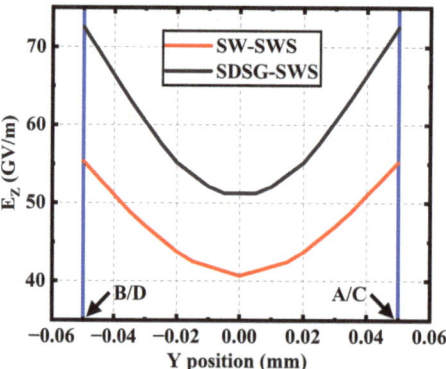

**Figure 8.** Variation of $E_Z$ in the y-o-z plane along the y-direction (A–B and C–D).

The vacuum model of the SDSG-SWS with couplers is shown in Figure 9a, which is mainly composed of a beam tunnel, the main slow-wave circuit, the mode converter, and the input–output waveguide. The main slow-wave circuit consists of 120 cycles. The operating mode of the SDSG-SWS is generally the $EH$ mode, while the mode of input–output waveguide is $TE_{10}$ mode. Therefore, the mode converter is designed to convert the $TE_{10}$ mode to the $EH$ mode in order to ensure that the input signal can be effectively coupled into the slow-wave circuit and stably amplified without reflection. In Figure 9b, it can be observed that the length of the mode converter is four periods, in which the height of gratings decreases proportionally towards the centerline of gratings until it becomes a smooth rectangular waveguide. As shown in Figure 9b, the electric field can gradually change from $EH$ mode to $TE_{10}$ mode through the coupler.

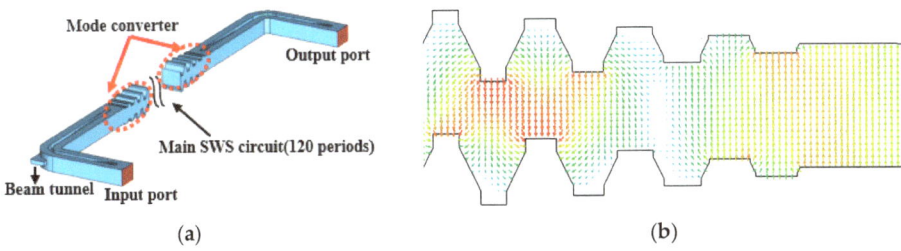

**Figure 9.** (**a**) Vacuum model of SDSG slow-wave circuit with couplers. (**b**) Cross-sectional view of the coupler electric field in the *y-o-z* plane.

According to the model shown in Figure 9a, the calculation results of electromagnetic transmission characteristics of the SDSG slow-wave circuit are shown in Figure 10. From 319 GHz to 438 GHz, $S_{11}$ is below $-17.9$ dB, while $S_{21}$ is above $-15$ dB.

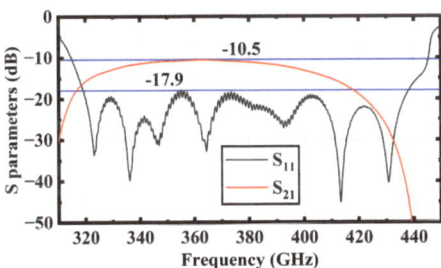

**Figure 10.** Transmission characteristics of SDSG slow-wave circuit with couplers.

## 3. Beam–Wave Interaction Simulation

The performance of the SDSG-TWT and the SW-TWT were analyzed using the PIC simulation of CST Particle Studio. In the PIC simulation, oxygen-free copper was used as the circuit material, and its conductivity is $1.8 \times 10^7$ S/m, considering the distribution loss of the circuit. In order to illustrate the advantages of TWTs using the SDSG-SWS in saturated power, gain, and electron efficiency, TWTs using the SDSG-SWS and SW-SWS should be kept at the same operating voltage and current. According to the dispersion characteristics shown in Figure 3, the synchronous operating voltage of both TWTs is set to 19.2 kV, and the operating current is set to 60 mA. Here, the tube length is assumed to be constant, and the output power is saturated by continuously increasing the input power. In CST, the grid number of the SDSG-TWT is set to 18,000,000 and the time required by the PC (2.9 GHz CPU and Tesla k20c accelerator card) is 25 h for a 12 ns simulation of a single input signal. The results are displayed in Figures 11–18.

Figure 11 shows the variation of signal amplitude over time at 340 GHz for TWTs using the SDSG-SWS. The results show that the SDSG-TWT reaches a stable amplification state after 0.8 ns and remains without oscillation. The SDSG-TWT achieves an output voltage of 8.1 V (corresponding power of 32.8 W) at an input voltage of 0.35 V (corresponding power of 0.06 W).

Figure 12 presents the energy distribution of electrons in the phase space along the longitudinal direction when the signal remains at stable amplification for a long time. The results show that there are more decelerating electrons than accelerating electrons. Most of the electronic energy is converted into the energy of the electromagnetic wave. It can be observed that the electromagnetic wave signal is amplified.

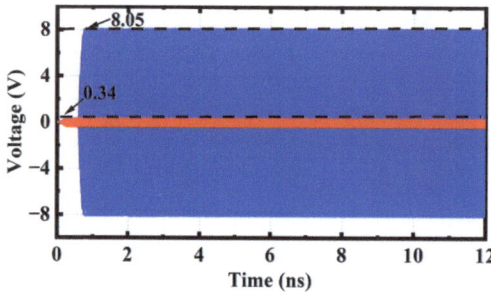

**Figure 11.** The signal amplitude versus time at 340 GHz.

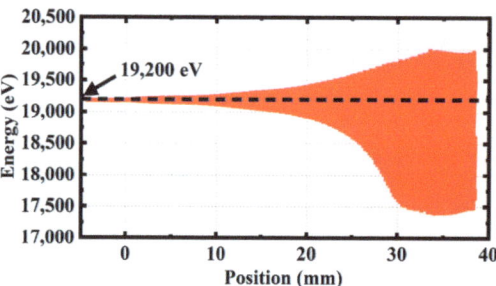

**Figure 12.** Phase momentum plot of the electrons.

Figure 13 is a full cycle electric field diagram. The results show that, as the longitudinal distance increases, the electric field intensity within the SWS also increases, which indirectly confirms that the SDSG-TWT can effectively amplify the input signal.

Figure 14 shows the longitudinal and transverse cross-sectional views of the electron beam (The cross-sectional view shows the connection between the SWS and the output coupler). The longitudinal cross-sectional view shows that electric field energy increases with increasing longitudinal distance. At the same time, near the end of the SWS circuit, electronic modulation reaches saturation. This result is consistent with the previous phase space diagram. The cross-sectional view shows that the electrons are not near the red line around them (The red line indicates the size of the electron beam channel). This result indicates that the modulated electrons were not intercepted by the metal wall.

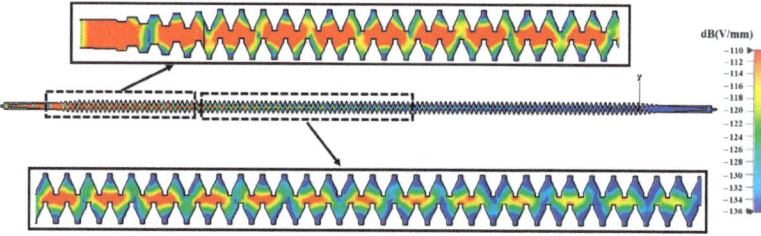

**Figure 13.** Electric field cross-section (*y*-direction).

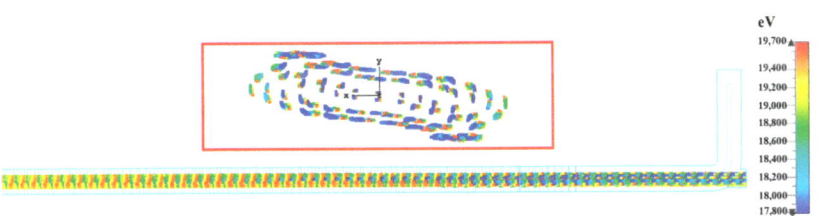

**Figure 14.** Transverse and longitudinal cross-sectional view of the electron trajectory.

Figure 15 shows the spectrum of the output signal. The Fourier transform of the output signal shows that that, with the exception of 340 GHz, the signals' amplitudes at other frequencies are extremely low, to the extent that they can be ignored. It indicates that the SDSG-TWT can effectively amplify the fundamental signal of 340 GHz without the oscillation starting of other signals.

Figures 16–18 show the performance comparison between the SDSG-TWT and the SW-TWT in terms of their saturated output power, gain, and electron efficiency. The results shown in Figures 16–18 indicate that the saturated output powers of the SDSG-TWT and SW-TWT are 32.8 W and 23.1 W; the 3 dB bandwidths are 316 GHz–405 GHz and 315 GHz–370 GHz; the maximum gains are 1.19 dB/mm and 0.61 dB/mm; and the maximum electron efficiencies are 2.84% and 1.80%, respectively. According to these results, it can be calculated that, compared with the SW-TWT, the SDSG-TWT demonstrates a 41% improvement in saturated output power, a 61.8% improvement in 3 dB bandwidth, an 83% improvement in gain, and a 63.3% improvement in electron efficiency under the same operating conditions.

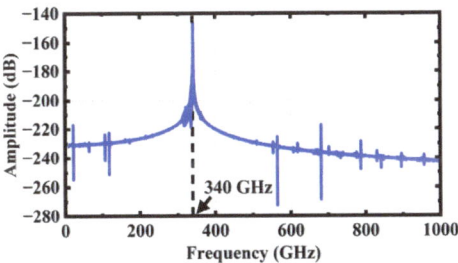

**Figure 15.** Frequency spectrum of output signal.

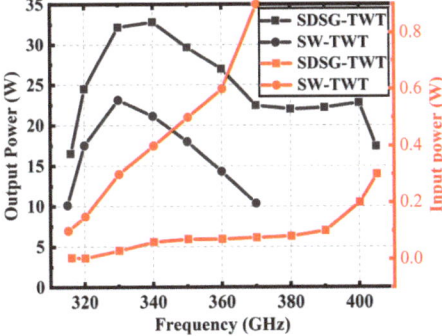

**Figure 16.** Output–input power versus frequency.

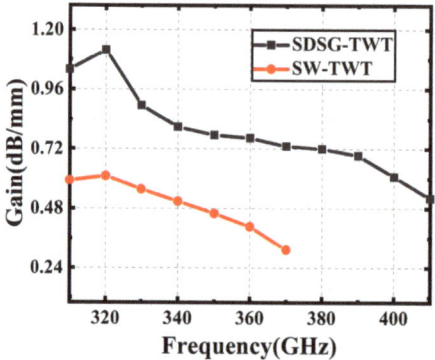

**Figure 17.** Gain versus frequency.

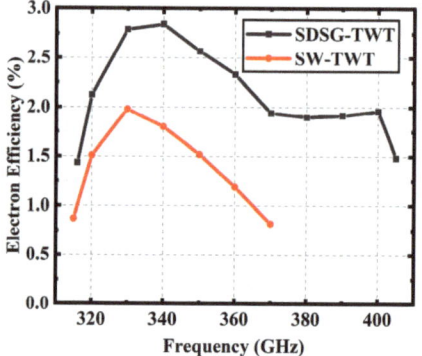

**Figure 18.** Electron efficiency versus frequency.

Table 2 shows a comparison of the performance between the proposed SDSG and three reported improved SWs. Compared with the new SW-SWS [21], the SDSG-SWS demonstrates significant advantages in operating voltage, operating current, gain, output power, and electronic efficiency, due to the MBSC-SWG-SWS [22] being a multi electron beam channel structure. Therefore, compared to the MBSC-SWG-SWS, the SDSG-SWS only has certain advantages in terms of gain. However, for 340 GHz TWT, it is very difficult to design the electron optics system of a multi-beam TWT. Therefore, the structure presented in this article is more applicable and practical. Compared with the modified SW-SWS [13], the SDSG-TWT has excellent performance in all aspects.

**Table 2.** Comparison of SDSG-SWS with the published SWS at 340 GHz.

| Structure | V (kV) | I (mA) | Outpower (W) | Gain (dB) | $\eta$ (%) |
|---|---|---|---|---|---|
| SDSG | 19.2 | 60 | 32.8 | 37.87 | 2.84 |
| New SW [21] | 12.65 | 30 | 10 | 20 | 2.63 |
| MBSC-SWG [22] | 21.3 | 54 | 51 | 24 | 4.43 |
| Modified SW [13] | 9.9 | 40 | 10 | 27 | 2.52 |

In summary, the SDSG-TWT demonstrates significant improvements in saturated output power, gain, and electron efficiency compared with the SW-TWT. PIC simulation results further validate the performance advantages of the SDSG-SWS over the SW-SWS.

## 4. Conclusions

A new SWS, called staggered double-segmented grating (SDSG), which is a combination of the SW-SWS and SDG-SWS, is investigated. Research has shown that it has the following characteristics: wide operating band, high interaction impedance, low loss, and ease of fabrication. Compared with the SW-TWT, the SDSG-TWT can produce higher output power, greater gain, and electron efficiency under the same operating conditions. Therefore, the SDSG-SWS can be regarded as a very promising submillimeter TWT slow-wave circuit.

**Author Contributions:** Conceptualization, J.Z. and Z.L.; methodology, Z.W. (Zechuan Wang) and J.Z.; software, J.Z. and Z.W. (Zechuan Wang); validation, J.D., H.C., S.W., Z.W. (Zhanliang Wang), and H.G.; formal analysis, J.Z. and Z.W. (Zechuan Wang); investigation, J.Z.; resources, Z.L. and Y.G.; data curation, J.Z., Z.W. (Zechuan Wang), and Z.L.; writing—original draft preparation, Z.W. (Zechuan Wang); writing—review and editing, J.Z., Z.L. and Y.G.; visualization, Z.W. (Zechuan Wang) and J.Z.; supervision, Z.W. (Zechuan Wang) and J.Z.; project administration, Z.W. (Zechuan Wang) and J.Z.; funding acquisition, Z.W. (Zechuan Wang) and J.Z. All authors have read and agreed to the published version of the manuscript.

**Funding:** This work was supported by the National Natural Science Foundation of China under Grants 62071087, 61921002, 92163204, 61988102 and 62150052, and supported by the Sichuan Science and Technology Program under Grant 2023NSFSC0452.

**Institutional Review Board Statement:** Not applicable.

**Informed Consent Statement:** Not applicable.

**Data Availability Statement:** Data sharing is not applicable.

**Conflicts of Interest:** The authors declare no conflict of interest.

## References

1. Siegel, P.H. Terahertz technology. *IEEE Trans. Microw. Theory Tech.* **2002**, *50*, 910–928. [CrossRef]
2. Booske, J.H.; Dobbs, R.J.; Joye, C.D.; Kory, C.L.; Neil, G.R.; Park, G.S.; Park, J.; Temkin, R.J. Vacuum Electronic High Power Terahertz Sources. *IEEE Trans. Terahertz Sci. Technol.* **2011**, *1*, 54–75. [CrossRef]
3. Hu, M.; Zhong, R.; Gong, S.; Zhao, T.; Liu, D.; Liu, S. Tunable Free-Electron-Driven Terahertz Diffraction Radiation Source. *IEEE Trans. Electron Devices* **2018**, *65*, 1151–1157. [CrossRef]
4. Sherwin, M. Terahertz power. *Nature* **2002**, *420*, 131–133. [CrossRef] [PubMed]
5. Cai, J.; Wu, X.; Feng, J. Traveling-Wave Tube Harmonic Amplifier in Terahertz and Experimental Demonstration. *IEEE Trans. Electron Devices* **2015**, *62*, 648–651.
6. Wang, W.; Zhang, Z.; Wang, P.; Zhao, Y.; Zhang, F.; Ruan, C. Double-mode and double-beam staggered double-vane traveling-wave tube with high-power and broadband at terahertz band. *Sci. Rep.* **2022**, *12*, 12012. [CrossRef]
7. Bhattacharjee, S.; Booske, J.H.; Kory, C.L.; Weide, D.W.v.d.; Limbach, S.; Gallagher, S.; Welter, J.D.; Lopez, M.R.; Gilgenbach, R.M.; Ives, R.L.; et al. Folded waveguide traveling-wave tube sources for terahertz radiation. *IEEE Trans. Plasma Sci.* **2004**, *32*, 1002–1014. [CrossRef]
8. Cai, J.; Feng, J.; Wu, X. Folded Waveguide Slow Wave Structure With Modified Circular Bends. *IEEE Trans. Electron Devices* **2014**, *61*, 3534–3538.
9. Tian, Y.; Yue, L.; Wang, H.; Zhou, Q.; Wei, Y.; Hao, B.; Wei, Y.; Gong, Y. Investigation of Ridge-Loaded Folded Rectangular Groove Waveguide Slow-Wave Structure for High-Power Terahertz TWT. *IEEE Trans. Electron Devices* **2018**, *65*, 2170–2176. [CrossRef]
10. Shin, Y.M.; Baig, A.; Barnett, L.R.; Luhmann, N.C.; Pasour, J.; Larsen, P. Modeling Investigation of an Ultrawideband Terahertz Sheet Beam Traveling-Wave Tube Amplifier Circuit. *IEEE Trans. Electron Devices* **2011**, *58*, 3213–3218. [CrossRef]
11. Shin, Y.M.; Baig, A.; Barnett, L.R.; Tsai, W.C.; Luhmann, N.C. System Design Analysis of a 0.22-THz Sheet-Beam Traveling-Wave Tube Amplifier. *IEEE Trans. Electron Devices* **2012**, *59*, 234–240. [CrossRef]
12. Zhu, J.; Lu, Z.; Duan, J.; Wang, Z.; Gong, H.; Gong, Y. A Modified Staggered Double Grating Slow Wave Structure for W-Band Dual-Beam TWT. *IEEE Trans. Electron Devices* **2023**, *70*, 320–326. [CrossRef]
13. Choi, W.; Lee, I.; Choi, E. Design and Fabrication of a 300 GHz Modified Sine Waveguide Traveling-Wave Tube Using a Nanocomputer Numerical Control Machine. *IEEE Trans. Electron Devices* **2017**, *64*, 2955–2962. [CrossRef]
14. Fang, S.; Xu, J.; Hairong, Y.; Yin, P.; Lei, X.; Wu, G.; Yang, R.; Luo, J.; Yue, L.; Zhao, G.; et al. Design and Cold Test of Flat-Roofed Sine Waveguide Circuit for W-Band Traveling-Wave Tube. *IEEE Trans. Plasma Sci.* **2020**, *48*, 4021–4028. [CrossRef]
15. Zhang, L.; Jiang, Y.; Lei, W.; Hu, P.; Guo, J.; Song, R.; Tang, X.; Ma, G.; Chen, H.; Wei, Y. A piecewise sine waveguide for terahertz traveling wave tube. *Sci. Rep.* **2022**, *12*, 10449. [CrossRef]

16. Carlsten, B.E.; Russell, S.J.; Earley, L.M.; Krawczyk, F.L.; Potter, J.M.; Ferguson, P.; Humphries, S. Technology development for a mm-wave sheet-beam traveling-wave tube. *IEEE Trans. Plasma Sci.* **2005**, *33*, 85–93. [CrossRef]
17. Panda, P.C.; Srivastava, V.; Vohra, A. Analysis of Sheet Electron Beam Transport Under Uniform Magnetic Field. *IEEE Trans. Plasma Sci.* **2013**, *41*, 461–469. [CrossRef]
18. Su, Y.; Wang, P.; Wang, W.; Ruan, C.; He, W. Theoretical Analysis of Sheet Beam Electron Gun for Terahertz Vacuum Electron Devices. *IEEE Trans. Electron Devices* **2022**, *69*, 5865–5870. [CrossRef]
19. Lu, Z.; Zhu, M.; Ding, K.; Wen, R.; Ge, W.; Wang, Z.; Tang, T.; Gong, H.; Gong, Y. Investigation of Double Tunnel Sine Waveguide Slow-Wave Structure for Terahertz Dual-Beam TWT. *IEEE Trans. Electron Devices* **2020**, *67*, 2176–2181. [CrossRef]
20. Zhang, L.; Wei, Y.; Guo, G.; Ding, C.; Wang, Y.; Jiang, X.; Zhao, G.; Xu, J.; Wang, W.; Gong, Y. A Ridge-Loaded Sine Waveguide for $G$-Band Traveling-Wave Tube. *IEEE Trans. Plasma Sci.* **2016**, *44*, 2832–2837. [CrossRef]
21. Zhang, X.; Xu, J.; Fang, S.; Jiang, X.; Yin, P.; Luo, J.; Hu, Y.; Ge, X.; Yin, H.; Yue, L.; et al. A New type of 0.34THz Sine Waveguide Slow Wave Structure. In Proceedings of the 2020 IEEE 21st International Conference on Vacuum Electronics (IVEC), Monterey, CA, USA, 19–22 October 2020; pp. 233–234.
22. Luo, J.; Xu, J.; Yin, P.; Yang, R.; Yue, L.; Wang, Z.; Xu, L.; Feng, J.; Liu, W.; Wei, Y. A 340 GHz High-Power Multi-Beam Overmoded Flat-Roofed Sine Waveguide Traveling Wave Tube. *Electronics* **2021**, *10*, 3018. [CrossRef]

**Disclaimer/Publisher's Note:** The statements, opinions and data contained in all publications are solely those of the individual author(s) and contributor(s) and not of MDPI and/or the editor(s). MDPI and/or the editor(s) disclaim responsibility for any injury to people or property resulting from any ideas, methods, instructions or products referred to in the content.

*Communication*

# An Angular Radial Extended Interaction Amplifier at the W Band

Yang Dong, Shaomeng Wang *, Jingyu Guo, Zhanliang Wang, Huarong Gong, Zhigang Lu, Zhaoyun Duan and Yubin Gong

National Key Laboratory of Science and Technology on Vacuum Electronics, School of Electronic Science and Engineering, University of Electronic Science and Technology of China, No. 2006 Xiyuan Avenue, High-Tech District (West District), Chengdu 611731, China
* Correspondence: wangsm@uestc.edu.cn

**Abstract:** In this paper, an angular radial extended interaction amplifier (AREIA) that consists of a pair of angular extended interaction cavities is proposed. Both the convergence angle cavity and the divergence angle cavity, which are designed for the converging beam and diverging beam, respectively, are investigated to present the potential of the proposed AREIA. They are proposed and explored to improve the beam–wave interaction capability of W-band extended interaction klystrons (EIKs). Compared to conventional radial cavities, the angular cavities have greatly decreased the ohmic loss area and increased the characteristic impedance. Compared to the sheet beam (0°) cavity, it has been found that the convergence angle cavity has a higher effective impedance and the diverging beam has a weaker space-charge effect under the same ideal electron beam area; the advantages become more obvious as the propagation distance increases. Particle-in-cell (PIC) results have shown that the diverging beam (8°) EIA performs better at an output power of 94 GHz under the condition of lossless, while the converging beam (−2°) EIA has a higher output power of 6.24 kW under the conditions of ohmic loss, an input power of 0.5 W, and an ideal electron beam of 20.5 kV and 1.5 A. When the loss increases and the beam current decreases, the output power of the −2° EIA can be improved by nearly 30% compared to the 0° EIA, and the −2° EIA has a greatly improved beam–wave interaction capacity than conventional EIAs under those conditions. In addition, an angular radial electron gun is designed.

**Keywords:** AREIA; convergence angle; effective impedance; space-charge effect

## 1. Introduction

Radial vacuum electron devices (RVEDs) were first proposed by Arman in 1994, in which the electron beam diverged along the radial direction. Additionally, he also researched radial oscillators [1] and accelerators [2], which can obtain high power in the low-frequency band under relativistic conditions. In [3–6], theoretical, simulational, and experimental research was conducted on RVEDs, most of which operate under relativistic conditions and in the frequency band between the L band and Ku band. The above research shows that radial beam devices have the advantages of a low space-charge effect, a higher power capacity, and stronger beam–wave interactions compared to conventional devices.

However, the existing studies on radial klystron amplifiers are mainly focused on low-frequency bands and relativistic conditions. When the operating frequency is increased to the W band and the conditions are non-relativistic, the main problems are the modulation capability and ohmic loss of the radial cavity. Additionally, the radial cavity has a low *R/Q* and a large ohmic loss area. These problems can be ameliorated by the angular radial cavity, as shown in Figure 1c. The radial cavity size is greatly reduced by the divergence angle cavity. On the one hand, the characteristic impedance can be increased, and, on the other hand, the ohmic loss area is reduced.

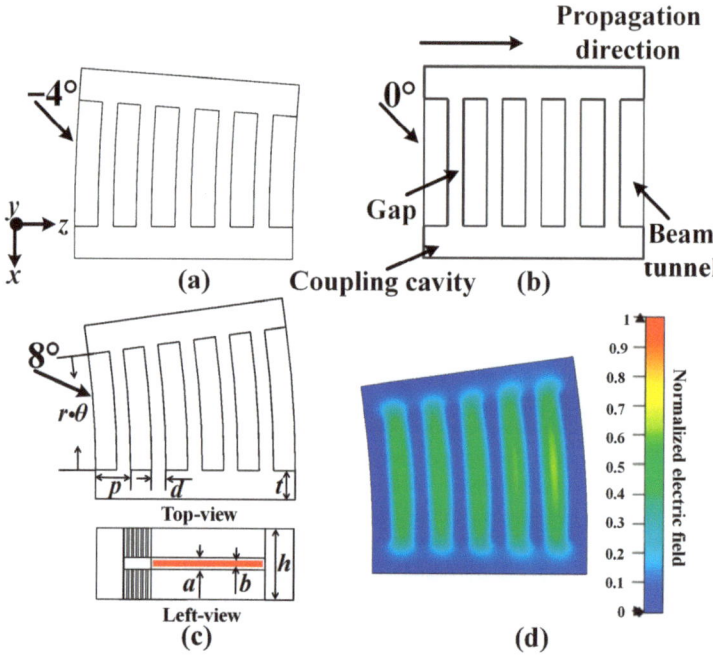

**Figure 1.** The structure of an AREIA, (**a**) −4° (convergence angle), (**b**) 0° (sheet beam), (**c**) 8° (divergence angle), (**d**) E-field distribution of the $2\pi$ mode.

Considering the current condition of the low-beam, the limiting of the space-charge effect is decreased, and the modulation capability becomes more important. To improve the modulation capability of the angular radial cavity in that condition, the convergence angle cavity is proposed to operate a converging beam, as shown in Figure 1a. With the propagation of the converging beam, the characteristic impedance of the corresponding convergence angle cavity can be increased.

In W-band or terahertz (THz) klystrons, the extended interaction cavity is widely used, which has the advantages of a large power capacity and a high characteristic impedance. Related research includes W-band pencil beam EIAs [7–9], W-band multi-beam EIAs [10], W-band sheet beam EIAs [11], G-band sheet beam EIAs [12,13], G-band pencil beam EIAs [14,15], and W-band tested EIKs [16–19]. In general, the sheet beam has a more uniform beam–wave interaction than the pencil beam, and the current density of the beam can be greatly decreased by the sheet beam. For W-band sheet beam EIAs, the relevant studies are significantly fewer than those for pencil beam EIAs.

In reference [17], the proposed EIK with three extended interaction cavities is driven by a sheet beam of 20 kV and 4 A, which can produce 7.5 kW of peak power with a beam–wave interaction efficiency of 9.38%, and it adopts a single-stage depressed collector and water cooling. In references [7–9], the operating currents of the pencil beam are less than 1 A, and their beam–wave interaction efficiencies are both less than 10%. To improve the beam–wave interaction capability of W-band EIKs, the AREIAs are implemented and explored.

In this paper, the diverging beam EIA and converging beam EIA will be analyzed from the perspective of beam–wave interactions and compared with the sheet beam EIA (SEIA), which can be seen as a special angular beam EIA with an angle of 0°. In Section 2, the design and dispersion of the cavities are analyzed, and the effective impedance ($M^2*(R/Q)$) of the different cavities and the space-charge effect of the different beams are studied. In Section 3, the hot performance of the AREIA with three cavities is simulated. In Section 4, an angular radial electron gun of −2° is designed.

## 2. Design and Analysis

The angular radial extended interaction cavities with five gaps are shown in Figure 1. As the distance of the propagation increases, the current density of the beam in the convergence angle cavity (Figure 1a) will increase, but it will decrease in the divergence angle cavity (Figure 1c). If the angle is set to zero, the angular radial cavity will turn out to be a sheet beam cavity (Figure 1b). The $2\pi$ mode (Figure 1d), which is suitable for the input and output of the signal, is chosen as the operation mode. Additionally, the optimized dimension parameters of the cavity are shown in Table 1. Figure 2 shows the dispersion diagrams of $-4°$, $0°$, and $8°$ cavities with the dimensions from Table 1, and they are obtained in the CST Eigenmode Solver. The synchronous voltages of the $7\pi/4$ and $9\pi/4$ modes are 29.3 kV and 17.1 kV, respectively; thus, those adjacent modes will not compete with the $2\pi$ mode.

**Table 1.** AREIA dimensions.

| Symbol | Quantity | Dimension (mm) |
|---|---|---|
| $r$ | Initial radius | 20 \| 40 \| 40 |
| $\theta$ | Angular angle | 8° \| 4° \| −4° |
| $p$ | Length of period | 0.88 |
| $d$ | Gap width | 0.35 |
| $a$ | Beam tunnel thickness | 0.30 |
| $b$ | Electron beam thickness | 0.20 |
| $h$ | Height of gap | 1.72 |
| $t$ | Coupling cavity width | 0.70 |

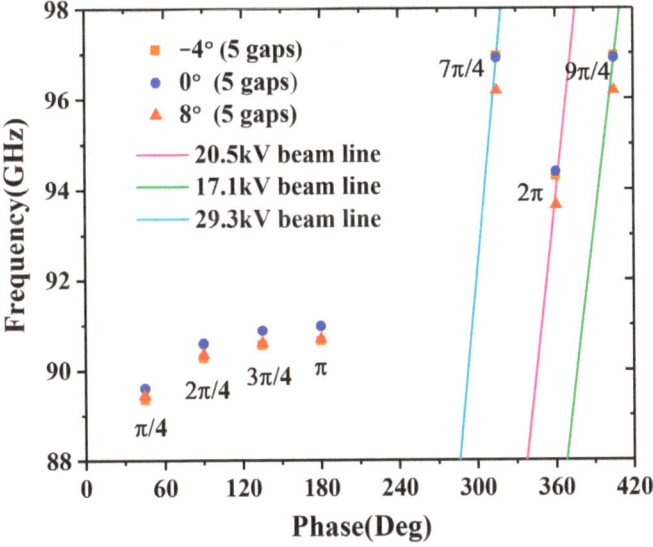

**Figure 2.** Dispersion diagrams of the different angle cavities.

To evaluate the interaction capability of the different angle EIAs, the effects of three main parameters, i.e., $M^2*(R/Q)$, space-charge field, and ohmic loss, are investigated for the different angle cavities. Assuming that the cavities have the same initial arc length (2.79 mm) at $\Delta r = 0$ mm during the investigation, the characteristic impedance ($R/Q$) and coupling coefficient ($M$) can be calculated as follows [14]:

$$R/Q = \frac{\left(\int_{-\infty}^{+\infty}|E_z|dz\right)^2}{2wW_s}, M = \frac{\int_{-\infty}^{+\infty}E_z e^{j\beta_e z}dz}{\int_{-\infty}^{+\infty}|E_z|dz}. \quad (1)$$

where $W_s$ is the total energy storage, $w$ is the angular frequency, $E_z$ is the axial electric field, and $\beta_e$ is the propagation constant of the dc beam.

The higher the $M^{2*}(R/Q)$ of the cavity, the higher the degree of electron beam modulation; however, the space-charge field will prevent this process. The $M^{2*}(R/Q)$ of the $-4°$ cavity will increase in propagation, but the $8°$ cavity does the opposite, as shown in Figure 3. When $\Delta r = 0$ mm, there is little difference in $M^{2*}(R/Q)$ between the $-4°$, $0°$, and $8°$ cavities, which means that they have almost the same modulation capability in the input cavity. In summary, the convergence angle cavity has stronger modulation capability than other cavities, and the advantage will increase as the propagation distance increases.

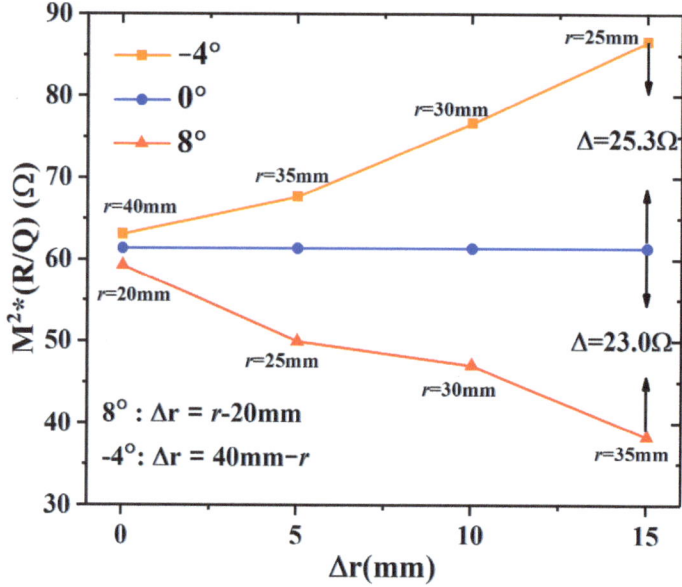

**Figure 3.** The effective impedances of the different angle cavities vary with the different initial radii in the propagation.

In the Cartesian coordinate system, Green's function in the region of $[-l/2, l/2] \times [-a/2, a/2] \times (-\infty, +\infty)$ can be written as follows [20]:

$$G(M, M_0) = \frac{2}{al\varepsilon_0} \sum_{m=0}^{\infty} \sum_{n=0}^{\infty} \frac{e^{-k_z|z-z_0|}}{k_z} \cos(k_x x) \cos(k_x x_0) \cos(k_y y) \cos(k_y y_0),$$
$$k_x = \frac{(2n+1)\pi}{l}, k_y = \frac{(2m+1)\pi}{a}, k_z^2 = k_x^2 + k_y^2. \quad (2)$$

In the discrete particle case, the charge of each particle is $-q$ ($q = I/f/N$, where $I$ is the beam current and $N$ is the number of particles in one period), and then the axial space-charge field of each particle can be calculated as follows:

$$\begin{aligned}|E_{scz}(M_0)| &= \left|\frac{\partial}{\partial z} \sum_{i=1}^{N} qG(M_i, M_0)\right| \\ &= \frac{2q}{al\varepsilon_0} \sum_{i=1}^{N} \sum_{m=0}^{\infty} \sum_{n=0}^{\infty} e^{-k_z|z_i-z_0|} \cos(k_x x_i) \cos(k_x x_0) \cos(k_y y_i) \cos(k_y y_0).\end{aligned} \quad (3)$$

It can be seen that $|E_{scz}|$ is positively correlated with $\cos(k_x x_i)$ and inversely correlated with $|z_i - z_0|$. The diverging beam has a larger range of $|x_i|$, but the converging beam does the opposite, which means that the former has the advantage of a weaker space-charge effect. Since $|z_i-z_0|$ depends on the degree of bunching, the stronger the bunching, the stronger the space-charge effect.

Table 2 lists the variations of $Q_0$ ($Q_0 = wW_s/P_{loss}$, where $P_{loss}$ is the loss power) for the $-4°$, $0°$, and $8°$ cavities with different conductivities, and there is little difference between the cavities at the same conductivity. This indicates that the ohmic loss power differs slightly between them under the same energy storage conditions. The ohmic loss power can also be written as follows:

$$P_{loss} = \iint_S \frac{1}{2}|J_s|^2 R_s dS = \iint_S \frac{1}{2}|J_s|^2 \sqrt{\frac{\pi \mu f}{\sigma}} dS. \quad (4)$$

where $J_s$ is the surface current density, $\mu$ is the magnetic permeability, and $\sigma$ denotes the conductivity. It can be found that the loss is related to the interaction area, surface current density, operating frequency, and conductivity. With the same surface current density, the divergence angle cavity will lose more power than the convergence angle cavity.

**Table 2.** $Q_0$ with different conductivities.

| $\sigma$ ($\times 10^7$ S/m) | $-4°$ ($r$ = 25 mm) | $0°$ | $8°$ ($r$ = 35 mm) |
|---|---|---|---|
| 5.8 | 1315 | 1184 | 1253 |
| 3.6 | 1036 | 933 | 987 |
| 2.2 | 810 | 729 | 771 |

## 3. Simulation Performance

Figure 4 shows the metal models of three different angle EIAs, all with three five-gap cavities; the standard waveguide WR-10 (2.54 mm × 1.27 mm) is used as the input and output port. A PIC simulation is conducted by CST Particle Studio; a uniform hexahedral mesh is used (the number of cells for the $-2°$ AREIA is 2,147,740), and the boundary conditions are all set as ideal electrical boundaries.

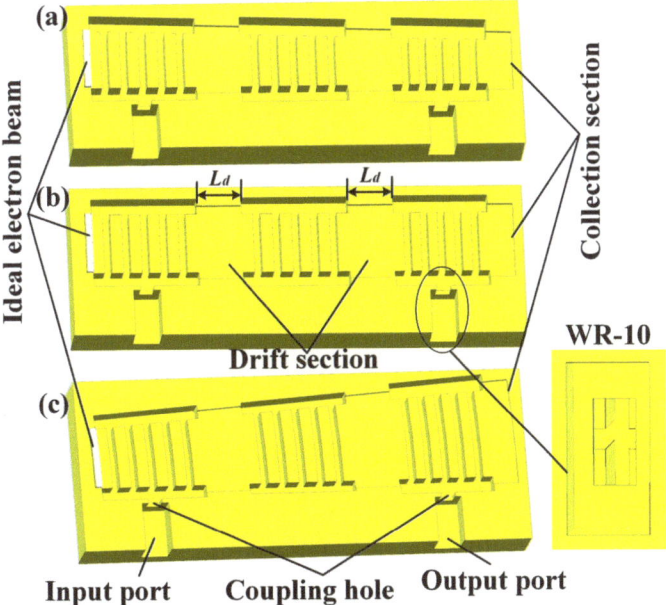

**Figure 4.** A schematic diagram (half-metal model) of the three types of EIAs: (**a**) $-2°$, (**b**) $0°$, and (**c**) $4°$.

For the comparison, three measures are adopted. First, although the ideal beams of the three EIAs are different, they have the same ideal beam area. The ideal beam current density (269 A/cm$^2$) of each EIA is set to the same, and a radial focusing magnetic field of 0.8 T is used to ensure that all of the AREIAs can be focused. Second, the input cavity has the same $Q_e$, and the coupling hole of the output cavity is optimized for the optimal output power of each EIA. Third, since the loading of the electron beam will affect the resonant frequency, the frequency of each cavity has been fine-tuned to work best at 94 GHz by adjusting $t$. Additionally, the drift distance ($L_d$ = 2.07 mm) between the cavities and the total interaction length (20 mm) of the different EIAs are the same, and the period length of the output cavity ($p_{out}$) is set to 0.80 mm; thus, the electron beam can deliver more energy.

For the PEC (perfect electrical conductor) shielding case, the voltage and current of the electron beam are 20.5 kV and 1.5 A, respectively, and the input power is 0.1 W. The PIC results are shown in Figure 5, and the output power increases with the angle. When $\theta = 8°$, the output power of 7.45 kW increased by 6.1% compared with the 0° EIA. The modulations of the input cavities are almost no different, as shown in Figure 6. In the output cavity, the modulation of the 8° EIA is stronger than that of the 0° EIA and −2° EIA, whether in the acceleration zone or the deceleration zone. This indicates that the diverging beam EIA has taken advantage of the weaker space-charge effect, but the advantage is not great under these conditions.

For the lossy metal shielding case, the metal material is set as copper, with a conductivity of $5.8 \times 10^7$ S/m, and the input power is set to 0.5 W. The output power and gain reach their maximum values (6.24 kW and 41.0 dB, respectively) at −2°, as shown in Figure 7, and the output power increases by 5.1% compared with the 0° EIA. The output power of the −4° EIA is lower than that of the −2° EIA due to the stronger space-charge effect. Instead, the output power of the diverging beam EIA drops due to the larger loss area, although it has taken advantage of the weaker space-charge effect.

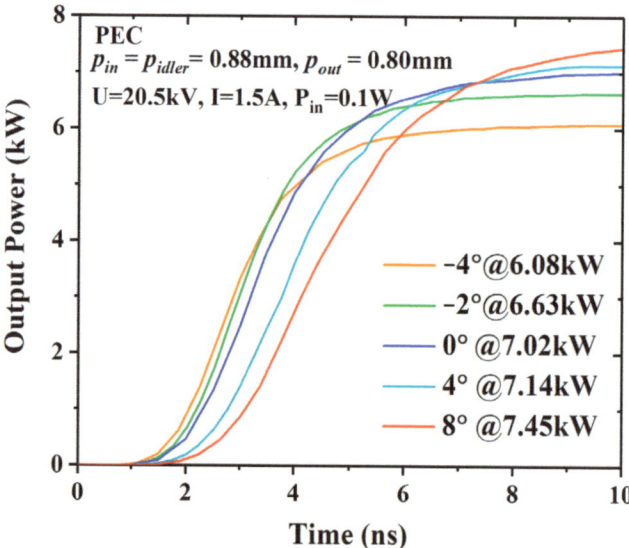

**Figure 5.** The output powers of the different EIAs in the case of lossless.

The output powers of the −2°, 0°, and 4° EIAs with the different input powers are shown in Figure 8, and the saturated output powers are 6.59 kW, 6.06 kW, and 5.64 kW, respectively. The saturated output power of the −2° EIA increases by 8.7% compared with the 0° EIA. Figure 9 shows that the output power of the −2° EIA, 0° EIA, and 4° EIA varies with the frequency under the input power of 0.5 W, and the 3 dB bandwidths are 285 MHz,

270 MHz, and 230 MHz, respectively. The 3 dB bandwidth has little difference between the −2° EIA and 0° EIA, but the difference is up to 55 MHz between the −2° EIA and 4° EIA. The reason for this is that when the operating frequency deviates, the interaction degree decreases and the space-charge restriction effect is weakened, meaning that the stronger cavity modulation makes the −2° EIA bandwidth wider than that of the 4° EIA under the same interaction length and input power.

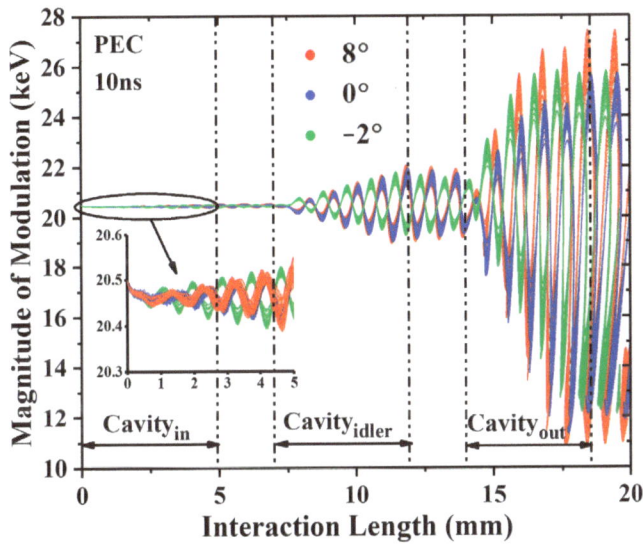

**Figure 6.** The phase space diagrams of the −2°, 0°, and 8° EIAs in the case of lossless.

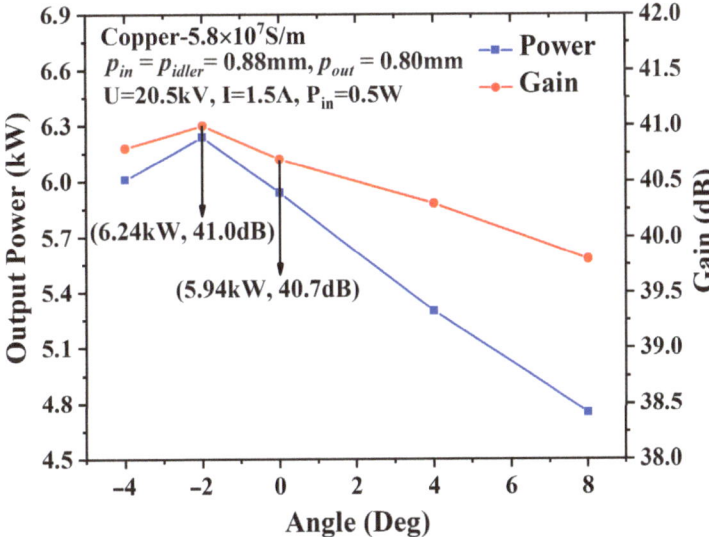

**Figure 7.** The output power and gain of the different angles in the case of loss.

The influence of loss on the output power is further analyzed in Table 3. When $\sigma$ decreases from $5.8 \times 10^7$ S/m to $2.2 \times 10^7$ S/m, the differences in the output power between the −2° EIA and the 0° EIA are 0.3 kW, 0.3 kW, and 0.37 kW. Additionally, the difference between the 0° EIA and 4° EIA is more obvious.

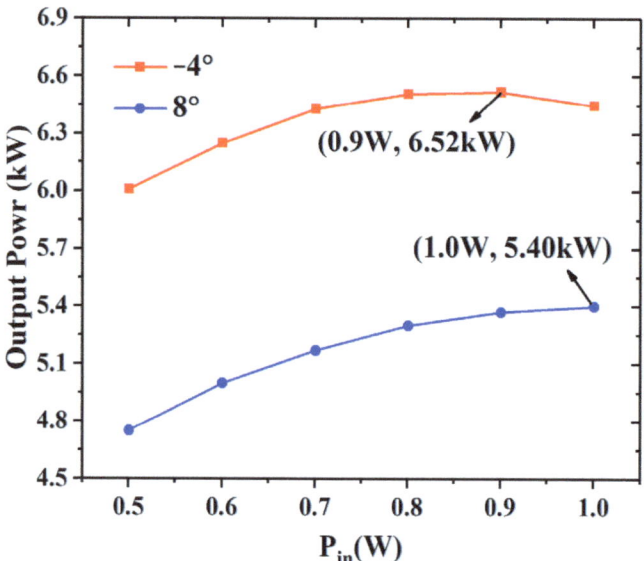

**Figure 8.** The output power of the different EIAs varies with the input power in the case of loss.

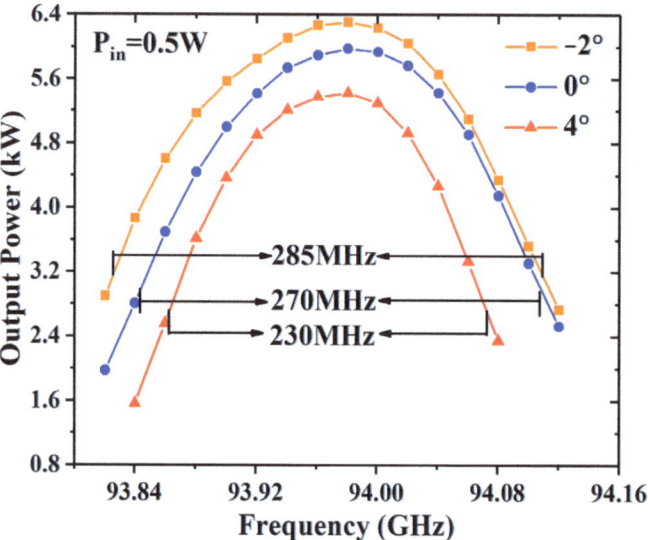

**Figure 9.** The output power of the different EIAs varies with the frequency in the case of loss.

**Table 3.** Output powers with different conductivities.

| σ (×$10^7$ S/m) | −2° EIA (kW) | 0° EIA (kW) | 4° EIA (kW) |
| --- | --- | --- | --- |
| 5.8 | 6.24 | 5.94 | 5.30 |
| 3.6 | 5.46 | 5.16 | 4.41 |
| 2.2 | 4.42 | 4.05 | 3.28 |

When the beam current decreases to 0.8 A (an ideal beam current density of 143 A/cm$^2$), the space-charge effect is further depressed. As the input power increases from 0.5 W to 1 W, the difference in the output power between the $-2°$ EIA and the $0°$ EIA increases from 0.41 kW to 0.46 kW, as shown in Figure 10. When the input power is 1 W, the output power of the $-2°$ EIA increases by 29.7% compared with the $0°$ EIA.

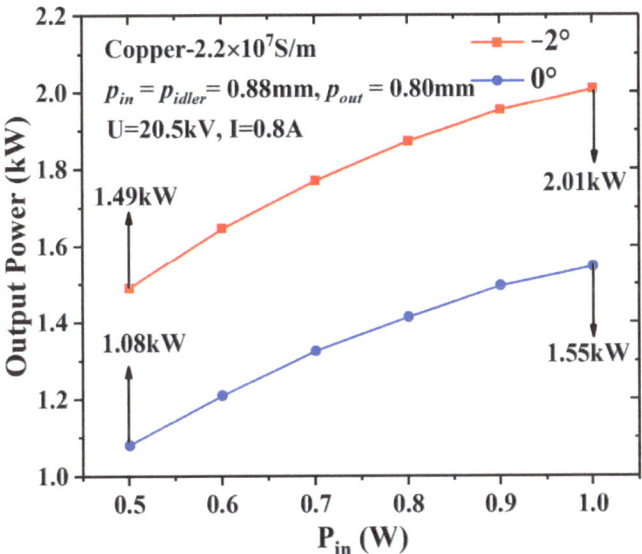

**Figure 10.** The output power of the different EIAs varies with the input power when the current is 0.8 A and the conductivity is $2.2 \times 10^7$ S/m.

Compared to previous studies [7,9,11,17], the sheet beam EIAs have a higher beam–wave interaction efficiency than the pencil beam EIAs, and the $-2°$ EIA has better interaction performance, as shown in Table 4. Even under the high loss condition, the beam–wave interaction efficiency of the $-2°$ EIA reached 12.3%, which is a 2.8% increase compared with $0°$ EIA.

**Table 4.** Comparison of the different EIAs.

| Type | Operating Parameters | Output Power & Efficiency |
| --- | --- | --- |
| Pencil beam EIA [7] | 5 kV, 0.2 A, $5.8 \times 10^7$ S/m | 67 W, 6.7% |
| Pencil beam EIA [9] | 16 kV, 0.6 A, $5.8 \times 10^7$ S/m | 0.9 kW, 9.4% |
| Sheet beam EIA [11] | 75 kV, 4 A, $5.8 \times 10^7$ S/m | 50 kW, 16.7% |
| Sheet beam EIA [17] | 20 kV, 4 A | 7.5 kW (peak, tested), 9.4% |
| $-2°$ EIA | 20.5 kV, 1.5 A, $5.8 \times 10^7$ S/m | 6.59 kW, 21.4% |
| $-2°$ EIA | 20.5 kV, 0.8 A, $2.2 \times 10^7$ S/m | 2.01 kW, 12.3% |
| $0°$ EIA | 20.5 kV, 0.8 A, $2.2 \times 10^7$ S/m | 1.55 kW, 9.5% |

In Figure 11, the magnitude of modulation will drop with the decrease in conductivity and beam current, especially the beam current. As it can be seen from Table 3 and Figure 11, with the increase in the loss or decrease in the beam current, the bunching is decreased, which means that the charge density in the bunching center is reduced. Additionally, the space-charge effect is depressed, as known from (3); thus, the converging beam EIA has a greater advantage. Since the balance between the space-charge effect and the cavity modulation changes with the operating conditions and structure dimensions, the optimal angle also changes.

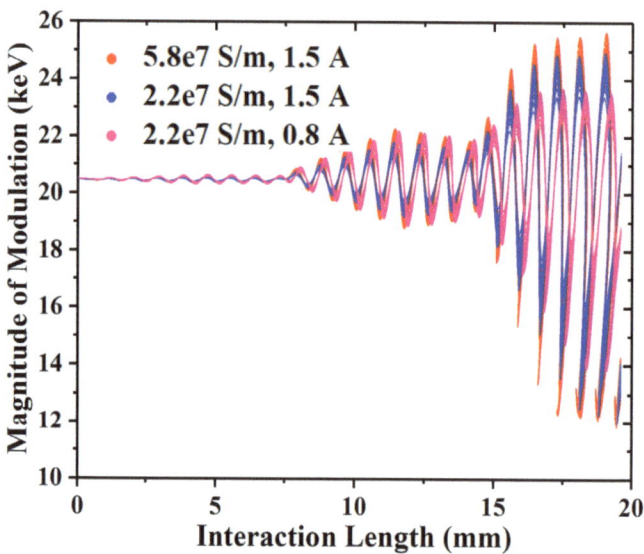

**Figure 11.** The phase space diagrams of the −2° EIA with different conductivities and beam currents.

## 4. Design of the Angular Radial Electron Gun

The divergence angle electron optical system has been designed in [21], and Figure 12 shows the sketch of the −2° angular radial electron gun. In this case, the −2° AREIA has an advantage; the beam voltage and the current are set to 20.5 kV and 0.8 A (a current density of 143 A/cm$^2$), respectively. Considering the self-compression effect of the converging beam, the main requirement is to compress the beam thickness, as shown in Figure 12c. A beam size in the beam–wave entrance of 80 mm × 2° × 0.2 mm, a cathode emission area of 88 mm × 2° × 1.5 mm (a corresponding emission current density of 17.4 A/cm$^2$), and an external radial focusing magnetic field of 0.5 T are added in the beam–wave section. Figure 13 shows that the beam can be compressed well under self-compression, which is obtained in the CST Particle Tracking Solver. The compressed beam size meets the requirements at the beam–wave interaction entrance, as shown in Figure 13a. Additionally, in Figure 13b, the beam transmission is 100%.

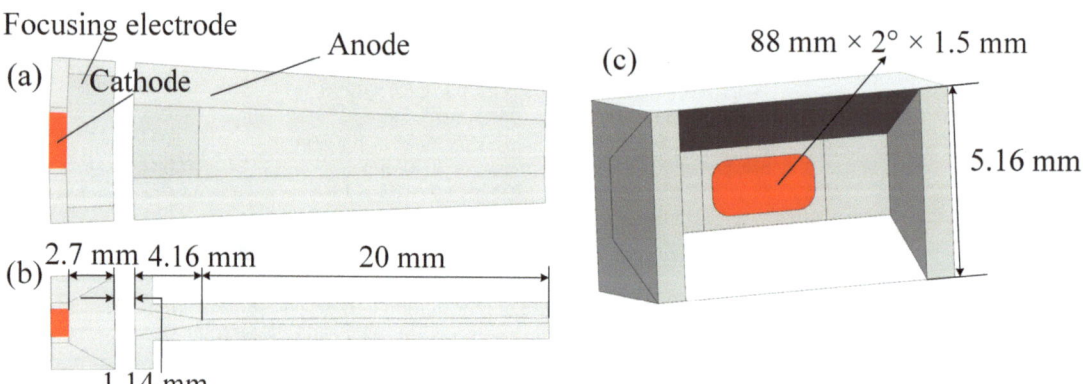

**Figure 12.** A schematic diagram of the −2° electron gun: (**a**) x–z plane cross section; (**b**) y–z plane cross section; (**c**) focusing electrode and cathode.

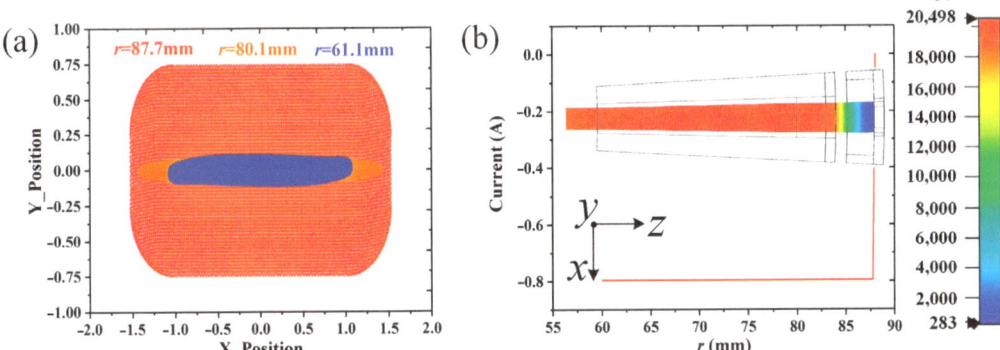

**Figure 13.** (**a**) The cross-sectional distribution of the particles, and (**b**) the total emission current along the propagation.

## 5. Conclusions

In our previous work [22], a 0.14 THz angular extended interaction oscillator with a diverging beam was studied and verified. The fabricated angular cavity had little machining error, which is also suitable for the processing of the $-2°$ cavity. In this paper, the convergence angle cavity and divergence angle cavity are proposed to operate the converging beam EIA and diverging beam EIA, respectively.

The convergence angle cavity is proved to have a stronger modulation capability than the sheet beam cavity, and the diverging beam has a weaker space-charge effect than the sheet beam. The diverging beam EIA can have an advantage in output power under stronger beam–wave interactions; in the case of lossy metals, the $-2°$ EIA manages to provide the maximum output power of 94 GHz, and the advantage becomes more obvious as the beam–wave interaction degree decreases, which is affected by the ohmic loss and beam current. The detailed research above can provide an important reference for the study of angular radial devices, such as angular radial traveling-wave tubes and angular radial backward-wave tubes.

The concept of the converging beam EIA can be further developed at the W band or the THz band. The $\pi$ mode operation has been used in sheet beam EIAs [13] and pencil beam EIAs [8,14], and the angular radial beam is also suitable for $\pi$ mode operations. The angular radial beam EIA is suitable for angular radial integration, which can be used for multi-beam operations. In future work, processing and cold tests will be conducted. The divergence angle electron optical system has been designed in reference [21], and the design and assembly of convergence angle electron optical systems require further research and implementation.

**Author Contributions:** Conceptualization, Y.D. and S.W.; methodology, Y.D. and J.G.; software, Y.D. and J.G.; validation, S.W., Z.W., H.G. and Z.L.; formal analysis, Y.D. and J.G.; investigation, Y.D.; resources, S.W., Z.D. and Y.G.; data curation, Y.D. and S.W.; writing—original draft preparation, Y.D.; writing—review and editing, S.W. and Y.G.; visualization, Y.D.; supervision, Y.G.; project administration, Y.G.; funding acquisition, Y.G. All authors have read and agreed to the published version of the manuscript.

**Funding:** This work is supported by the National Natural Science Foundation of China (Grant Nos. 61921002, 61988102, 92163204, 62071087, and 62150052).

**Institutional Review Board Statement:** Not applicable.

**Informed Consent Statement:** Not applicable.

**Data Availability Statement:** Data sharing is not applicable.

**Conflicts of Interest:** The authors declare no conflict of interest.

## References

1. Arman, M.J. High-power radial klystron oscillator. *Proc. SPIE* **1995**, *2557*, 21–31.
2. Arman, M.J. Radial acceletron, a new low-impedance HPM source. *IEEE Trans. Plasma Sci.* **1996**, *24*, 964–969. [CrossRef]
3. Wu, Z.F.; Wang, Y.Z. Theoretical design and numerical simulations of radial klystron oscillator. *HPL Part. Beams* **2000**, *12*, 211–214.
4. Li, S.F.; Ding, W. A high power microwave radial klystron oscillator with foldaway concentric cylinder resonant cavity. *HPL Part. Beams* **2003**, *15*, 909–913.
5. Zang, J.F.; Liu, Q.X.; Lin, Y.C.; Zhu, J. High frequency characteristics of radial three-cavity transmit time oscillator. *HPL Part. Beams* **2008**, *20*, 2046–2050.
6. Dang, F.C.; Zhang, X.P.; Zhong, H.H.; Zhang, J.; Ju, J.C. A high efficiency Ku-band radial line relativistic klystron amplifier. *Phys. Plasmas* **2016**, *23*, 073113.
7. Chang, Z.W.; Meng, L.; Yin, Y.; Wang, B.; Li, H.L.; Rauf, A.; Ullah, S.; Bi, L.J.; Peng, R.B. Circuit Design of a Compact 5-kV W-Band Extended Interaction Klystron. *IEEE Trans. Electron Dev.* **2018**, *65*, 1179–1184. [CrossRef]
8. Chang, Z.; Shu, G.; Tian, Y.; He, W. Study of the π-Mode Operation in the Extended Interaction Circuit. *IEEE Trans. Plasma Sci.* **2022**, *50*, 649–655. [CrossRef]
9. Naidu, V.B.; Gope, D.; Datta, S.K. Enhancement of Bandwidth of an Extended Interaction Klystron by Symmetric Loading. *IEEE Trans. Plasma Sci.* **2022**, *50*, 5018–5022.
10. Lü, S.; Zhang, C.; Wang, S.; Wang, Y. Stability Analysis of a Planar Multiple-Beam Circuit for a W-Band High-Power Extended-Interaction Klystron. *IEEE Trans. Electron Dev.* **2015**, *62*, 3042–3048.
11. Chen, S.Y.; Ruan, C.J.; Wang, Y.; Zhang, C.Q.; Zhao, D.; Yang, X.D.; Wang, S.Z. Particle-in-Cell Simulation and Optimization of Multigap Extended Output Cavity for a W-Band Sheet-Beam EIK. *IEEE Trans. Plasma Sci.* **2014**, *42*, 91–98.
12. Wang, H.; Xue, Q.; Zhao, D.; Qu, Z.; Ding, H. A Wideband Double-Sheet-Beam Extended Interaction Klystron with Ridge-Loaded Structure. *IEEE Trans. Plasma Sci.* **2022**, *50*, 1796–1802. [CrossRef]
13. Li, R.J.; Ruan, C.J.; Li, S.S.; Zhang, H.F. G-band Rectangular Beam Extended Interaction Klystron Based on Bi-Periodic Structure. *IEEE Trans. Terahertz Sci. Tech.* **2019**, *9*, 498–504. [CrossRef]
14. Li, S.S.; Ruan, C.J.; Fahad, A.K.; Wang, P.P.; Zhang, Z.; He, W.L. Novel Coupling Cavities for Improving the Performance of G-Band Ladder-Type Multigap Extended Interaction Klystrons. *IEEE Trans. Plasma Sci.* **2020**, *48*, 1350–1356. [CrossRef]
15. Hyttinen, M.; Roitman, A.; Horoyski, P.; Deng, H. High Power Pulsed 263 GHz Extended Interaction Amplifier. In Proceedings of the 2020 IEEE 21st International Conference on Vacuum Electronics (IVEC), Monterey, CA, USA, 19–22 October 2020.
16. Pasour, J.; Wright, E.; Nguyen, K.; Levush, B. Compact, multi-kW sheet beam oscillator at 94 GHz. In Proceedings of the 2014 IEEE 41st International Conference on Plasma Sciences (ICOPS), Washington, DC, USA, 25–29 May 2014.
17. Pasour, J.; Wright, E.; Nguyen, K.T.; Balkcum, A.; Wood, F.N.; Myers, R.E. Demonstration of a Multikilowatt, Solenoidally Focused Sheet Beam Amplifier at 94 GHz. *IEEE Trans. Electron Dev.* **2014**, *61*, 1630–1636. [CrossRef]
18. Berry, D.; Deng, H.; Dobbs, R.; Horoyski, P.; Hyttinen, M.; Kingsmill, A.; MacHattie, R.; Roitman, A.; Sokol, E.; Steer, B. Practical Aspects of EIK Technology. *IEEE Trans. Electron Dev.* **2014**, *61*, 1830–1835. [CrossRef]
19. Wei, Y.; Li, D.; Zhou, J.; Yang, J.; Yin, L.; Ouyang, J. A High Power W-band Extended Interaction Klystron. In Proceedings of the 2019 International Vacuum Electronics Conference (IVEC), Busan, Republic of Korea, 29 April–1 May 2019.
20. Rowe, J.E. *Nonlinear Electron-Wave Interaction Phenomena*; Academic Press: Cambridge, MA, USA, 1965; pp. 72–77.
21. Li, X.Y.; Wang, Z.L.; He, T.L.; Duan, Z.Y.; Wei, Y.Y.; Gong, Y.B. Study on Radial Sheet Beam Electron Optical System for Miniature Low-Voltage Traveling-Wave Tube. *IEEE Trans. Electron Dev.* **2017**, *64*, 3405–3412. [CrossRef]
22. Dong, Y.; Wang, S.M.; Guo, J.Y.; Wang, Z.L.; Tang, T.; Gong, H.R.; Lu, Z.G.; Duan, Z.Y.; Gong, Y.B. A 0.14 THz Angular Radial Extended Interaction Oscillator. *IEEE Trans. Electron Dev.* **2022**, *69*, 1468–1473.

**Disclaimer/Publisher's Note:** The statements, opinions and data contained in all publications are solely those of the individual author(s) and contributor(s) and not of MDPI and/or the editor(s). MDPI and/or the editor(s) disclaim responsibility for any injury to people or property resulting from any ideas, methods, instructions or products referred to in the content.

*Communication*

# Ultrafast Modulation of THz Waves Based on MoTe₂-Covered Metasurface

Xing Xu [1,2,3,4,†], Jing Lou [2,†], Mingxin Gao [2], Shiyou Wu [1,3,4], Guangyou Fang [1,3,4] and Yindong Huang [2,*]

[1] Aerospace Information Research Institute, Chinese Academy of Sciences, Beijing 100094, China
[2] Innovation Laboratory of Terahertz Biophysics, National Innovation Institute of Defense Technology, Beijing 100071, China
[3] Key Laboratory of Electromagnetic Radiation and Sensing Technology, Chinese Academy of Sciences, Beijing 100190, China
[4] School of Electronic, Electrical and Communication Engineering, University of Chinese Academy of Sciences, Beijing 100049, China
* Correspondence: yindonghuang@nudt.edu.cn
† These authors contributed equally to this work.

**Abstract:** The sixth generation (6G) communication will use the terahertz (THz) frequency band, which requires flexible regulation of THz waves. For the conventional metallic metasurface, its electromagnetic properties are hard to be changed once after being fabricated. To enrich the modulation of THz waves, we report an all-optically controlled reconfigurable electromagnetically induced transparency (EIT) effect in the hybrid metasurface integrated with a 10-nm thick MoTe₂ film. The experimental results demonstrate that under the excitation of the 800 nm femtosecond laser pulse with pump fluence of 3200 $\mu J/cm^2$, the modulation depth of THz transmission amplitude at the EIT window can reach 77%. Moreover, a group delay variation up to 4.6 ps is observed to indicate an actively tunable slow light behavior. The suppression and recovery of the EIT resonance can be accomplished within sub-nanoseconds, enabling an ultrafast THz photo-switching and providing a promising candidate for the on-chip devices of the upcoming 6G communication.

**Keywords:** photo-switching; terahertz; reconfigurable metasurface; ultrafast dynamics

## 1. Introduction

Located in the electromagnetic spectrum between the microwave and infrared region, the terahertz (THz) frequency band is a spectral window with great scientific interests [1–3]. Nowadays, the THz technology has found wide applications, including nondestructive sensing [4], biomedicine [5], security inspection [6], and communication [7–9]. As one of the most promising applications, the THz wireless communication has attracted tremendous attention in promoting the development of the sixth generation (6G) communication network, and is expected to enable the implementation of the "internet of everything" in the near future [10–12]. Generally, THz radiation can offer higher carrier frequency and spatial resolution than the microwave does. Meanwhile, it can penetrate a large number of the non-polar non-metallic substances, such as the silicon, plastic, clothing and paper, opening the way of novel communication to be used by satellites, autonomous cars, smart cities and so on [13–15]. It requires a series of regulations to THz waves when using as the 6G communication carriers. However, the limited availability of THz materials in nature prevents the flexible modulation of THz radiation [16].

The appearance of metasurfaces has provided an effective solution to modulate THz waves [17–19]. By altering the geometry and arrangement pattern of sub-wavelength meta-atoms in metasurfaces, the amplitude, phase and polarization of THz waves can be engineered, presenting numerous unusual applications, such as ultrathin flat lenses [20,21], THz broadband filter [22], and THz vortex beam generation [23,24]. Generally, once a planar metal structure is fabricated, its electromagnetic (EM) properties are unable to be actively tuned, which blocks the further flexible regulation of EM waves. To solve this problem, the concept of reconfigurable metasurfaces comprised of the active media and meta-atoms has been proposed [25–27]. By introducing the external excitation such as the optical [28–33], electrical [34–36], thermal [37–39] and mechanical stimuli [40,41], the intrinsic properties or physical form of the active media can be changed, affecting the EM properties of the metasurface. For example, for the THz asymmetric split ring resonators (TASRs) covered by a 310-nm thick germanium (Ge) film, photogenerated carriers can be excited by the 800 nm pump laser stimulus to change the resonance state [32]. Specifically, without pump laser excitation, the TASRs exhibits Fano resonance for the incident polarized THz radiation, denoted as the "on" state. When the pump laser reaches a threshold energy, the Fano resonance can be quenched, denoted as the "off" state. Moreover, compared with the electrical, thermal, and mechanical stimuli, the optical excitation possesses the unique advantage of ultrafast response speed. This means that the switch between "on" and "off" states can be achieved within picoseconds to nanoseconds. It shows the potential of providing an ultrafast data processing speed that is applicable in the 6G communication.

To realize the ultrafast switchable resonance state, one key point is to utilize the active semiconductor materials. Photocarrier dynamics in the active materials directly affect the switching time and modulation depth of the resonance. Nowadays, two-dimensional transition metal dichalcogenides (TMDCs) have received much attention due to their excellent photoelectric properties, such as the adjustable bandgap with the layer, high carrier mobility and good stability in the atmosphere [42,43]. $MoTe_2$ is a significant TMDC semiconductor. The monolayer $MoTe_2$ possesses a direct bandgap of about 1.1 eV [44], and the bulk form exhibits an indirect bandgap of about 0.88 eV [45]. Such a small bandgap enables a large number of photogenerated carriers in the material under the pump of the commercially available 800 nm laser (photon energy of 1.55 eV). Moreover, the carrier mobility of $MoTe_2$ can reach 8.5 $cm^2/V/s$ at the thickness of 10 nm [46]. These outstanding characteristics promote the application of $MoTe_2$ in photodetectors [47], phototransistors [48], field-effect transistors [49], and sensors [50]. However, as a prominent TMDC semiconductor, the application of $MoTe_2$ in the ultrafast THz switch has not been reported.

In this paper, we report an ultrafast switchable transmission amplitude modulation as well as the slow light behavior by integrating the 10-nm thick 2H-type $MoTe_2$ film with the electromagnetically induced transparency (EIT) metasurface, for the first time. Based on the homemade optical pump and THz probe (OPTP) spectroscopy system, the EIT amplitude modulation depth reaches 77% under the pump fluence of 3200 $\mu J/cm^2$ and shows a trend of quenching with higher pump energy. Moreover, a group delay variation up to 4.6 ps is measured to characterize the tunable slow light performance. It is demonstrated that the whole switch cycle from the "off" to "on" state of the EIT resonance can be completed on a timescale of the sub-nanosecond, providing a useful option for the ultrafast switchable THz metadevice that may be employed in the upcoming 6G communication.

## 2. Materials and Methods

### 2.1. MoTe$_2$-Covered EIT Metasurface

Figure 1a shows the schematic illustration of the switchable EIT metasurface device. Specifically, functional meta-atoms composed of gold H-shaped cut wires (HW) and parallel cut wires (PW) are periodically arranged to realize the EIT resonance of the incident THz waves. The detailed geometric configuration of each meta-atom is presented in Figure 1b. The metasurface was fabricated on the sapphire substrate using the conventional photolithography technology, with its optical microscopic image shown in Figure 1c. Then, the 10-nm thick 2H-type $MoTe_2$ film produced by the chemical vapor deposition was transferred onto the entire metasurface via the wet-transfer method [51], to complete the fabrication of the device. The surface morphology of the $MoTe_2$-coated metasurface can be clearly observed in Figure 1d. When an external 800 nm femtosecond laser pulse is irradiated on the $MoTe_2$ film, a large number of electrons can be excited to affect the coupling of the bright element HW and dark element PW, leading to the quenching of the EIT resonance. Immediately after that, the free electrons will recombine with the holes, and the EIT resonance will recover. The excitation and relaxation of such photogenerated carriers can be accomplished on the timescale of sub-nanosecond, enabling the ultrafast switching of the EIT resonant states. Generally, the external laser pulse is denoted as the optical pump, and the incident THz wave is denoted as the THz probe [32,52,53].

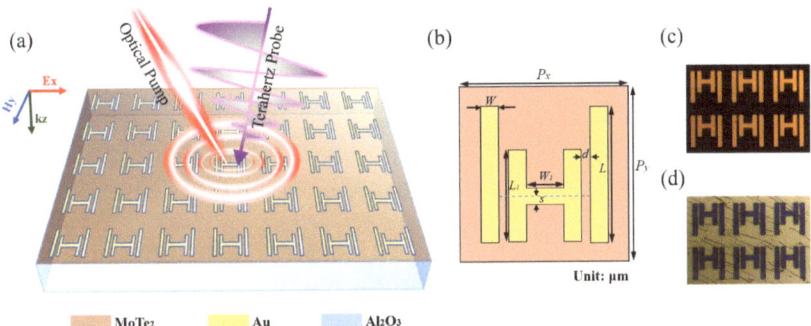

**Figure 1.** Schematic and working principle of the designed switchable EIT metasurface covered by the 10-nm thick MoTe$_2$ multilayer. (**a**) An artistic illustration of the hybrid metasurface under illumination of the 800 nm femtosecond laser and THz probe pulse; (**b**) the geometrical configuration of one proposed unit cell with the following structure parameters: $Px$ = 56 µm, $Py$ = 58.8 µm, $W$ = 5.6 µm, $L$ = 42 µm, $W_1$ = 12.6 µm, $L_1$ = 28 µm, $d$ = 3.5 µm, and $s$ = 2.8 µm. (**c**,**d**) Optical microscope images of the fabricated metasurface, without MoTe$_2$ and with 10-nm thick MoTe$_2$ coating, respectively.

### 2.2. Optical Pump and Terahertz Probe (OPTP) Measurement

To characterize the ultrafast switchable EIT resonance of the device, we have built an optical pump and THz probe (OPTP) spectroscopy system, as illustrated in Figure 2. A Ti: sapphire amplifier of Spectra-Physics was adopted as the laser source, with parameters of the 800 nm central wavelength, 100 fs pulse duration, 5 mJ pulse energy and 1 kHz repetition rate. The laser pulse was split into three beams for the optical pump (Beam 1), THz generation (Beam 2), and electro-optic sampling (EOS) (Beam 3). In the light path of optical pump, Beam 1 firstly passed through a half-wave plate (HW) and a thin film polarizer (TFP). By rotating the HW, the intensity of the laser transmitted from the TFP can be adjusted. Then, after passing the delay line 1 (DL1) and penetrating the indium tin oxide transparent conductive film glass (ITO), the optical pump irradiated onto the surface of the hybrid metasurface. In another light path, Beam 2 was focused by a lens (focal length of 30 cm) and then acted on the zinc telluride electro-optic crystal (ZnTe) to generate the THz emission via optical rectification. The emitted THz wave was collected by a pair of off-axis parabolic mirrors (OMP1 and OMP2), and the high-resistivity silicon plate (HR-Si) between the OMPs was used to block the 800 nm laser and transmit the THz emission. It is worth noting that the ITO can transmit the infrared wave and reflect the THz wave. Therefore, the optical pump from Beam 1 and the THz probe from Beam 2 can act on the hybrid metasurface jointly. To ensure a fully adequate optical excitation, the spot size of the optical pump beam on the sample was arranged to be larger than the THz beam. Moreover, by moving the translation stage of DL1, the optical path difference between the optical pump and THz probe can be changed to adjust the pump-probe delay, which is the key point for the characterization of the ultrafast dynamics. Then, the modulated THz waveform can be recorded by the EOS measurement with Beam 3, and the corresponding THz spectrum can be obtained by the standard Fourier transformation.

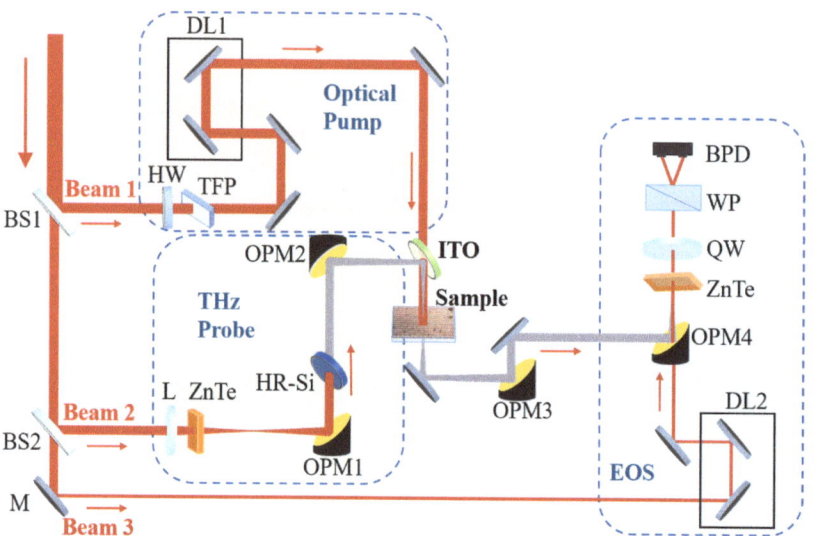

**Figure 2.** Schematic representation of the experimental setup. BS1 and BS2, beam splitters; M, mirror; L, lens; ZnTe, zinc telluride electro-optic crystal; OPM1-4, off-axis parabolic mirrors; HR-Si, high-resistivity silicon plate; ITO, indium tin oxide transparent conductive film glass; HW, half-wave plate; TFP, thin film polarizer; DL1 and DL2, delay lines; QW, quarter-wave plate; WP, Wollaston prism; BPD, balanced photodiodes.

## 3. Results

### 3.1. Simulation

In order to clarify the mechanism behind the modulation of MoTe$_2$-film-covered EIT metasurface, we have performed the numerical simulation by using the CST Microwave Studio Software in the time domain. The fabricated metasurface has a size of 10 mm × 10 mm. Here, we only need to simulate a unit cell of 56 μm × 58.8 μm size with the "Periodic" boundary conditions imposed in the $x$- and $y$-directions. The THz wave is set as a vertically incident plane wave polarized along the $x$ direction. A probe is placed at the bottom of the sapphire substrate to record the transmitted THz waveform.

As mentioned above, photogenerated carriers can be excited by the external laser pulse and will change the conductivity of the MoTe$_2$ film. Without laser pumping, we consider that the initial conductivity of the MoTe$_2$ film approximates to 0. For incident THz waves polarized along the $x$-direction, near-field coupling occurs between the H-shaped cut wire and parallel cut wires, leading to the high transmittance at 1.23 THz, as shown by the red line in Figure 3a. The corresponding near-field distribution is presented in Figure 3b, from which it can be observed that at the EIT peak frequency, the electric field of the H-shaped cut wire is suppressed, and the electromagnetic energy is transferred to the parallel cut wires. As we increase the conductivity of MoTe$_2$ film, the transmission amplitude of the EIT window gradually decreases, and finally, the EIT phenomenon disappears at the conductivity of $1 \times 10^5$ S/m (Figure 3a). Meanwhile, the near-field coupling gradually weakens so that less energy is transferred to the parallel cut wires, which is manifested as the large reduction of the electric field intensity above the metasurface, as shown in Figure 3c–e. These simulations help us to design and carry out the following experiments.

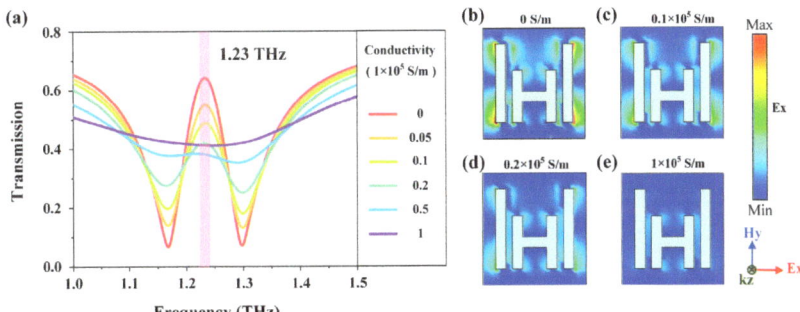

**Figure 3.** Simulated THz modulations of the hybrid EIT metasurface covered by the 10-nm thick MoTe$_2$ film. (**a**) THz transmission spectra with various conductivity of the MoTe$_2$ film from 0 S/m to $1 \times 10^5$ S/m. (**b**–**e**) Near E-field distributions above the metasurfaces at frequency of 1.23 THz within a unit cell.

## 3.2. Experimental Results

### 3.2.1. Pump-Fluence Controlled THz Transmission and Slow Light Behavior

Based on the homemade OPTP system, we have investigated the optical performance of the fabricated MoTe$_2$-coated metasurface. Firstly, the THz transmission spectra under various pump fluences were measured and compared to validate the feasibility of active modulation, as presented in Figure 4a. It can be found that the EIT window is suppressed significantly with increasing pump fluence, which is consistent with the simulation of the photoconductivity-dependent EIT resonance. The deviation of the transmission amplitude in the measurement might be introduced by the dielectric loss as well as the slight differences of the parameters and boundary conditions between the simulation model and the sample. To quantitatively evaluate the modulation effect, the modulation depth defined by the expression $MD_{EIT} = (T_0 - T_{pump})/T_0 \times 100\%$ is adopted [54]. In the expression, $T_0$ is the EIT amplitude without the pump and $T_{pump}$ is the EIT amplitude under the optical pump. The EIT amplitude is defined as the transmission amplitude difference between the peak (at 1.23 THz) and the valley (at 1.16 THz) of the EIT resonance [55].

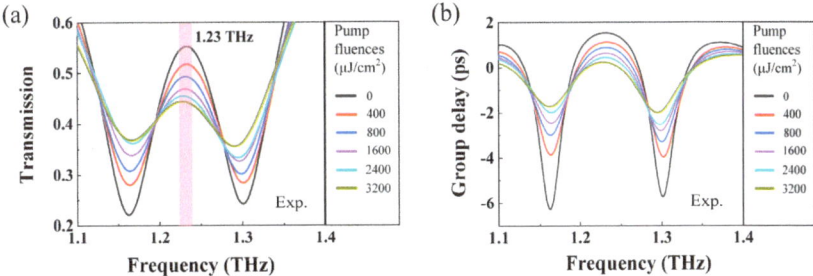

**Figure 4.** Pump fluence-dependent EIT response of the MoTe$_2$-coated metasurface. (**a**) Measured THz transmission spectra for different pump fluence from 0 to 3200 µJ/cm$^2$. (**b**) Measured group delays for different pump fluence from 0 to 3200 µJ/cm$^2$.

When the pump fluence reaches 3200 µJ/cm$^2$, the EIT modulation depth is calculated to be 77%, which is similar to the simulated modulation depth with the MoTe$_2$ conductivity of $2 \times 10^4$ S/m. Generally, photoconductivity is proportional to the product of the carrier mobility $\mu$ and the carrier density $n$, with the expression of $\sigma = \mu_e n_e e + \mu_h n_h q$. Based on the electron mobility of 8.5 cm$^2$/V/s in 10-nm thick MoTe$_2$ given by Ref [46], the carrier density under the pump fluence of 3200 µJ/cm$^2$ is estimated to be $1.47 \times 10^{26}$ m$^{-3}$, with the photoconductivity induced by the hole being ignored due to the fact that $\mu_e \gg \mu_h$. Here, 3200 µJ/cm$^2$ is the maximum pump fluence we can obtain in the experiment. If the pump laser energy can be further enhanced, it would be expected to achieve a larger modulation depth.

In addition to the pump fluence-dependent transmission spectra, the MoTe$_2$-coated metasurface also exhibits an actively controlled slow light behavior [56]. To characterize the slow light performance, the group delay defined as $\Delta t_g(\omega) = -d\phi/d\omega$ is employed, where $\phi$ is the relative phase of transmitted THz wave compared with that from the pure sapphire substrate, and $\omega$ is the angular frequency of the THz wave. The spectra of $\Delta t_g$ at different pump fluence is plotted in Figure 4b. Noted that the group delay reaches −6.3 ps at the EIT dip (1.16 THz) with no photo-injection, and increases to −1.7 ps at the maximum pump fluence. Such a high modulation depth of 4.6 ps can provide a wide manipulation range for light–matter interaction.

3.2.2. Ultrafast Dynamics of the Switching

In the last section, we have investigated the effect of pump fluence on THz transmission amplitude and group delay. It should be pointed out that the measurement was performed with a pump–probe delay of 0 ps to ensure the maximum modulation depth. The pump–probe delay is a variable that is used to represent the time difference to reach the metasurface between the THz pulse and the pump laser pulse. When the delay equals to 0, it implies the highest concentration of the photogenerated carriers when the THz probe interacts with the metasurface. The negative sign of the pump–probe delay means that the THz pulse reaches the metasurface earlier than the pump laser, whereas the positive sign represents the opposite.

Next, we will show the THz transmission spectra and group delay spectra at different pump-probe delays under the fixed pump fluence of 3200 μJ/cm$^2$, to manifest the ultrafast switch of the EIT response. Figure 5a,b show the switching-off process of the EIT resonance, which corresponds to the excitation of photogenerated carriers. Herein, the pump–probe delay starts from −10 ps. In that case, the pump laser pulse incident on the metasurface just met the 'tail' of the THz probe pulse, and the induced carriers do not affect the EIT resonance. As the pump–probe delay gradually increases to 0, the pump pulse starts to catch up with the main peak of the THz pulse, and more photogenerated carriers are concentrated to diminish the EIT response. After that, the switching-on processes are plotted in Figure 5c,d, with the EIT resonance displaying an explicit trend of recovery. It can be observed that the switching-on process of the EIT takes longer time than the switching-off process, depending on the excitation and relaxation of the carriers. For the 10-nm thick MoTe$_2$ film in our experiments, photogenerated carriers can be excited in a few picoseconds, but take hundreds of picoseconds for relaxation.

Finally, to better analyze the performance of the MoTe$_2$-covered ultrafast all-optical tuning metasurface, the parameters of relevant research based on the typical semiconductors [33,57] and another kind of TMDC [55] are listed in Table 1. All of the hybrid metasurfaces listed are in the EIT resonance mode and are pumped by the laser of the 800 nm wavelength. By comparison, it is found that the hybrid metasurface with 10-nm thick MoTe$_2$ film possesses the shorter switching time than that with the 500-nm-thick Si film [33], and its pump threshold is of the same order of magnitude as that with the 200-nm-thick Ge film [57]. Moreover, compared with the 40-nm thick WSe$_2$ [55], which also belongs to the TMDCs, the 10-nm thick MoTe$_2$ can achieve a greater modulation depth within the designed metasurface. Although the pump threshold of MoTe$_2$ is not dominant, it is important to note that the film thickness we applied is the thinnest. As silicon-based devices approach the limits of Moore's Law, processes below 14 nm are increasingly challenging. The quasi-two-dimensional MoTe$_2$ with the thickness of 10 nm in our work has demonstrated the potential to surpass the silicon-based device in response speed, providing a promising candidate for the on-chip device. Moreover, the comparison of the simulated photoconductivity implies that 10-nm-thick MoTe$_2$ might possess an ultrahigh optical conductivity, demonstrating competitiveness in the promising application of photoelectric devices.

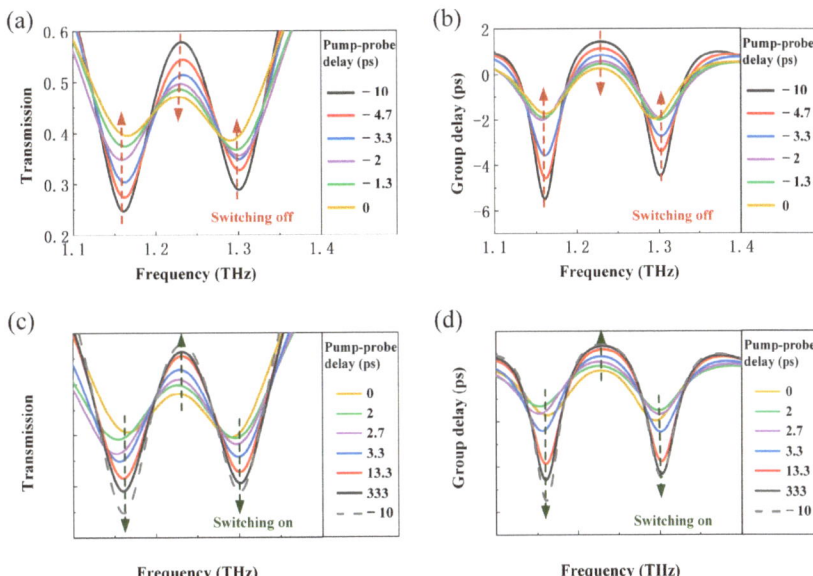

**Figure 5.** Ultrafast switching of the EIT resonance under pump fluence of 3200 µJ/cm$^2$. (**a**,**b**) The switching-off process of transmission spectra and group delay spectra as a function of pump–probe delay, respectively. (**c**,**d**) The switching-on process of transmission spectra and group delay spectra at various pump-probe delays, respectively.

**Table 1.** Comparison of the ultrafast all-optical tuning metasurfaces.

| Material | Thickness (nm) | Switching Time (ps) | Pump Threshold (µJ/cm$^2$) | Simulated Photoconductivity (S/m) | Modulation Depth |
|---|---|---|---|---|---|
| [33] Si | 500 | 780 (half-recovery state) | 200 | 600 | 100% |
| [57] Ge | 200 | 15 | 2200 | 1000 | 100% |
| [55] WSe$_2$ | 40 | 8 | 800 | 4800 | 43% |
| MoTe$_2$ | 10 | <300 (half-recovery state) | 3200 | >2 × 10$^4$ | 77% |

## 4. Conclusions and Perspective

In summary, we have demonstrated an all-optically controlled ultrafast switching of the THz transmission and slow light behavior by integrating the 2H-type MoTe$_2$ thin film with the EIT metasurface for the first time. The entire switching cycle can be completed in sub-nanoseconds, originating from the dynamic properties of the photocarriers in the MoTe$_2$ film. Based on the experimental system of OPTP, the THz transmission amplitude modulation up to 77% is achieved under a pump fluence of 3200 µJ/cm$^2$. In addition, a group delay variation as high as 4.6 ps is observed, indicating the slow light effect introduced by the fast change of refractive index on the resonant frequency of the designed EIT metasurface.

Our work provides a new candidate for on-chip devices on 6G communication, as well as an extraordinary slow light device to be applied in nonlinear optics, optical storage, and so on. The indirect band gap of MoTe$_2$ implies a possible application for near-infrared fiber communication located at the O-band (1260 nm–1360 nm) [58], enabling a hybrid control and ultrafast trigger of the infrared and THz waves. Furthermore, recent advances of the optically controlled THz metasurface have experimentally demonstrated a calibration-free sensor for achieving high-precision biosensing detection [31]. Since the biological interest concerns the full spectral range from THz wave to the mid-infrared (MIR) [59,60], the resonant frequency of the matasurface can also be extended to the MIR for more practical applications.

**Author Contributions:** Conceptualization, X.X., J.L., G.F. and Y.H; methodology, X.X., J.L. and Y.H; simulation, X.X. and J.L; investigation, X.X. and M.G.; writing—original draft preparation, X.X.; writing—review and editing, X.X., J.L., S.W, G.F. and Y.H; project administration, Y.H.; funding acquisition, Y.H, G.F. and S.W. All authors have read and agreed to the published version of the manuscript.

**Funding:** This research was funded by National Natural Science Foundation of China, Grant Nos. 12225511, T2241002, 12174449, 61988102 and 61971397.

**Institutional Review Board Statement:** Not applicable.

**Informed Consent Statement:** Not applicable.

**Data Availability Statement:** Data underlying the results presented in this paper are not publicly available at this time but may be obtained from the authors upon reasonable request.

**Conflicts of Interest:** The authors declare no conflict of interest.

# References

1. Zhang, X.C.; Shkurinov, A.; Zhang, Y. Extreme terahertz science. *Nat. Photonics* **2017**, *11*, 16–18. [CrossRef]
2. Salen, P.; Basini, M.; Bonetti, S.; Hebling, J.; Krasilnikov, M.; Nikitin, A.Y.; Shamuilov, G.; Tibai, Z.; Zhaunerchyk, V.; Goryashko, V. Matter manipulation with extreme terahertz light: Progress in the enabling THz technology. *Phys. Rep.-Rev. Sec. Phys. Lett.* **2019**, *836*, 1–74. [CrossRef]
3. Li, Y.; Chang, C.; Zhu, Z.; Sun, L.; Fan, C. Terahertz Wave Enhances Permeability of the Voltage-Gated Calcium Channel. *J. Am. Chem. Soc.* **2021**, *143*, 4311–4318. [CrossRef] [PubMed]
4. Tao, Y.H.; Fitzgerald, A.J.; Wallace, V.P. Non-Contact, Non-Destructive Testing in Various Industrial Sectors with Terahertz Technology. *Sensors* **2020**, *20*, 712. [CrossRef] [PubMed]
5. Xiang, Z.; Tang, C.; Chang, C.; Liu, G. A primary model of THz and far-infrared signal generation and conduction in neuron systems based on the hypothesis of the ordered phase of water molecules on the neuron surface I: Signal characteristics. *Sci. Bull.* **2020**, *65*, 308–317. [CrossRef]
6. Wang, L.; Yuan, M.; Huang, H.; Zhu, Y. Recognition of edge object of human body in THz security inspection system. *Infrared Laser Eng.* **2017**, *46*, 1125002. [CrossRef]
7. Yang, Y.; Yamagami, Y.; Yu, X.; Pitchappa, P.; Webber, J.; Zhang, B.; Fujita, M.; Nagatsuma, T.; Singh, R. Terahertz topological photonics for on-chip communication. *Nat. Photonics* **2020**, *14*, 446. [CrossRef]
8. Nagatsuma, T.; Ducournau, G.; Renaud, C.C. Advances in terahertz communications accelerated by photonics. *Nat. Photonics* **2016**, *10*, 371–379. [CrossRef]
9. Oleiwi, H.W.; Al-Raweshidy, H. SWIPT-Pairing Mechanism for Channel-Aware Cooperative H-NOMA in 6G Terahertz Communications. *Sensors* **2022**, *22*, 6200. [CrossRef] [PubMed]
10. Dang, S.; Amin, O.; Shihada, B.; Alouini, M.-S. What should 6G be? *Nat. Electron.* **2020**, *3*, 20–29. [CrossRef]
11. Saad, W.; Bennis, M.; Chen, M.Z. A Vision of 6G Wireless Systems: Applications, Trends, Technologies, and Open Research Problems. *IEEE Netw.* **2020**, *34*, 134–142. [CrossRef]
12. Wu, Z.H.; Zhang, T.R.; Li, Y.J.; Li, J.L.; Zhao, T.; Wang, W.; Song, T.; Liu, D.W.; Wei, Y.Y.; Gong, Y.B.; et al. Flexible terahertz dynamic tuning modulator: Top-to-bottom construction of in-plane gradient terahertz attenuator network. *Compos. Part A-Appl. Sci. Manuf.* **2022**, *163*, 107264. [CrossRef]
13. Shishanov, S.; Bystrov, A.; Hoare, E.G.; Stove, A.; Gashinova, M.; Cherniakov, M.; Tran, T.-Y.; Clarke, N. Height-Finding for Automotive THz Radars. *IEEE Trans. Intell. Transp. Syst.* **2019**, *20*, 1170–1180. [CrossRef]
14. Kokkoniemi, J.; Jornet, J.M.; Petrov, V.; Koucheryavy, Y.; Juntti, M. Channel Modeling and Performance Analysis of Airplane-Satellite Terahertz Band Communications. *IEEE Trans. Veh. Technol.* **2021**, *70*, 2047–2061. [CrossRef]
15. Farooq, M.S.; Nadir, R.M.; Rustam, F.; Hur, S.; Park, Y.; Ashraf, I. Nested Bee Hive: A Conceptual Multilayer Architecture for 6G in Futuristic Sustainable Smart Cities. *Sensors* **2022**, *22*, 5950. [CrossRef]
16. Ferguson, B.; Zhang, X.C. Materials for terahertz science and technology. *Nat. Mater.* **2002**, *1*, 26–33. [CrossRef]
17. Yu, N.; Genevet, P.; Kats, M.A.; Aieta, F.; Tetienne, J.P.; Capasso, F.; Gaburro, Z. Light propagation with phase discontinuities: Generalized laws of reflection and refraction. *Science* **2011**, *334*, 333–337. [CrossRef]
18. Sun, S.; He, Q.; Xiao, S.; Xu, Q.; Li, X.; Zhou, L. Gradient-index meta-surfaces as a bridge linking propagating waves and surface waves. *Nat. Mater.* **2012**, *11*, 426–431. [CrossRef]
19. Zang, X.; Yao, B.; Chen, L.; Xie, J.; Guo, X.; Balakin, A.V.; Shkurinov, A.P.; Zhuang, S. Metasurfaces for manipulating terahertz waves. *Light Adv. Manuf.* **2021**, *2*, 148–172. [CrossRef]
20. Wang, Q.; Zhang, X.; Xu, Y.; Tian, Z.; Gu, J.; Yue, W.; Zhang, S.; Han, J.; Zhang, W. A Broadband Metasurface-Based Terahertz Flat-Lens Array. *Adv. Opt. Mater.* **2015**, *3*, 779–785. [CrossRef]

21. Li, W.Y.; Sun, R.; Liu, J.Y.; Meng, T.H.; Zhao, G.Z. Broadband and high efficiency terahertz metasurfaces for anomalous refraction and vortex beam generation. *Chin. Phys. B* **2022**, *31*, 108701. [CrossRef]
22. Sun, D.D.; Qi, L.M.; Liu, Z.Y. Terahertz broadband filter and electromagnetically induced transparency structure with complementary metasurface. *Results Phys.* **2020**, *16*, 102887. [CrossRef]
23. Li, W.Y.; Zhao, G.Z.; Meng, T.H.; Sun, R.; Guo, J.Y. High efficiency and broad bandwidth terahertz vortex beam generation based on ultra-thin transmission Pancharatnam-Berry metasurfaces. *Chin. Phys. B* **2021**, *30*, 058103. [CrossRef]
24. Li, X.N.; Zhou, L.; Zhao, G.Z. Terahertz vortex beam generation based on reflective metasurface. *Acta Phys. Sin.* **2019**, *68*, 238101. [CrossRef]
25. Zheludev, N.I.; Kivshar, Y.S. From metamaterials to metadevices. *Nat. Mater.* **2012**, *11*, 917–924. [CrossRef]
26. He, Q.; Sun, S.; Zhou, L. Tunable/Reconfigurable Metasurfaces: Physics and Applications. *Research* **2019**, *2019*, 1849272. [CrossRef]
27. Fan, Y.C.; Zhao, Q.; Zhang, F.L.; Shen, N.H. Editorial: Tunable and Reconfigurable Optical Metamaterials. *Front. Phys.* **2021**, *9*, 713966. [CrossRef]
28. Gu, J.; Singh, R.; Liu, X.; Zhang, X.; Ma, Y.; Zhang, S.; Maier, S.A.; Tian, Z.; Azad, A.K.; Chen, H.-T.; et al. Active control of electromagnetically induced transparency analogue in terahertz metamaterials. *Nat. Commun.* **2012**, *3*, 1151. [CrossRef]
29. Cong, L.; Srivastava, Y.K.; Zhang, H.; Zhang, X.; Han, J.; Singh, R. All-optical active THz metasurfaces for ultrafast polarization switching and dynamic beam splitting. *Light-Sci. Appl.* **2018**, *7*, 28. [CrossRef]
30. Xu, X.; Lou, J.; Wu, S.; Yu, Y.; Liang, J.; Huang, Y.; Fang, G.; Chang, C. SnSe$_2$-functionalized ultrafast terahertz switch with ultralow pump threshold. *J. Mater. Chem. C* **2022**, *10*, 5805–5812. [CrossRef]
31. Lou, J.; Jiao, Y.; Yang, R.; Huang, Y.; Xu, X.; Zhang, L.; Ma, Z.; Yu, Y.; Peng, W.; Yuan, Y.; et al. Calibration-free, high-precision, and robust terahertz ultrafast metasurfaces for monitoring gastric cancers. *Proc. Natl. Acad. Sci. USA* **2022**, *119*, e2209218119. [CrossRef] [PubMed]
32. Lim, W.X.; Manjappa, M.; Srivastava, Y.K.; Cong, L.Q.; Kumar, A.; MacDonald, K.F.; Singh, R. Ultrafast All-Optical Switching of Germanium-Based Flexible Metaphotonic Devices. *Adv. Mater.* **2018**, *30*, 7. [CrossRef] [PubMed]
33. Hu, Y.Z.; You, J.; Tong, M.Y.; Zheng, X.; Xu, Z.J.; Cheng, X.G.; Jiang, T. Pump-Color Selective Control of Ultrafast All-Optical Switching Dynamics in Metaphotonic Devices. *Adv. Sci.* **2020**, *7*, 10. [CrossRef] [PubMed]
34. Lou, J.; Liang, J.; Yu, Y.; Ma, H.; Yang, R.; Fan, Y.; Wang, G.; Cai, T. Silicon-Based Terahertz Meta-Devices for Electrical Modulation of Fano Resonance and Transmission Amplitude. *Adv. Opt. Mater.* **2020**, *8*, 2000449. [CrossRef]
35. Tao, X.; Qi, L.M.; Fu, T.; Wang, B.; Uqaili, J.A.; Lan, C.W. A tunable dual-band asymmetric transmission metasurface with strong circular dichroism in the terahertz communication band. *Opt. Laser Technol.* **2022**, *150*, 107932. [CrossRef]
36. Nan, J.M.; Yang, R.S.; Xu, J.; Fu, Q.H.; Zhang, F.L.; Fan, Y.C. Actively modulated propagation of electromagnetic wave in hybrid metasurfaces containing graphene. *Epj Appl. Metamater.* **2021**, *7*, 9. [CrossRef]
37. Kim, Y.; Wu, P.C.; Sokhoyan, R.; Mauser, K.; Glaudell, R.; Shirmanesh, G.K.; Atwater, H.A. Phase Modulation with Electrically Tunable Vanadium Dioxide Phase-Change Metasurfaces. *Nano Lett.* **2019**, *19*, 3961–3968. [CrossRef]
38. Ma, H.; Wang, Y.; Lu, R.; Tan, F.; Fu, T.; Wang, G.; Wang, D.; Liu, K.; Fan, S.; Jiang, K.; et al. A flexible, multifunctional, active terahertz modulator with an ultra-low triggering threshold. *J. Mater. Chem. C* **2020**, *8*, 10213–10220. [CrossRef]
39. Liu, Z.Y.; Qi, L.M.; Lan, F.; Lan, C.W.; Yang, J.; Tao, X. A VO2 film-based multifunctional metasurface in the terahertz band. *Chin. Opt. Lett.* **2022**, *20*, 013602. [CrossRef]
40. Pitchappa, P.; Manjappa, M.; Ho, C.P.; Singh, R.; Singh, N.; Lee, C. Active Control of Electromagnetically Induced Transparency Analog in Terahertz MEMS Metamaterial. *Adv. Opt. Mater.* **2016**, *4*, 541–547. [CrossRef]
41. Zhu, W.M.; Liu, A.Q.; Bourouina, T.; Tsai, D.P.; Teng, J.H.; Zhang, X.H.; Lo, G.Q.; Kwong, D.L.; Zheludev, N.I. Microelectromechanical Maltese-cross metamaterial with tunable terahertz anisotropy. *Nat. Commun.* **2012**, *3*, 1274. [CrossRef] [PubMed]
42. Manzeli, S.; Ovchinnikov, D.; Pasquier, D.; Yazyev, O.V.; Kis, A. 2D transition metal dichalcogenides. *Nat. Rev. Mater.* **2017**, *2*, 17033. [CrossRef]
43. Pumera, M.; Sofer, Z.; Ambrosi, A. Layered transition metal dichalcogenides for electrochemical energy generation and storage. *J. Mater. Chem. A* **2014**, *2*, 8981–8987. [CrossRef]
44. Ruppert, C.; Aslan, O.B.; Heinz, T.F. Optical Properties and Band Gap of Single- and Few-Layer MoTe$_2$ Crystals. *Nano Lett.* **2014**, *14*, 6231–6236. [CrossRef]
45. Lezama, I.G.; Ubaldini, A.; Longobardi, M.; Giannini, E.; Renner, C.; Kuzmenko, A.B.; Morpurgo, A.F. Surface transport and band gap structure of exfoliated 2H-MoTe$_2$ crystals. *2D Mater.* **2014**, *1*, 021002. [CrossRef]
46. Ji, H.; Lee, G.; Joo, M.-K.; Yun, Y.; Yi, H.; Park, J.-H.; Suh, D.; Lim, S.C. Thickness-dependent carrier mobility of ambipolar MoTe$_2$: Interplay between interface trap and Coulomb scattering. *Appl. Phys. Lett.* **2017**, *110*, 183501. [CrossRef]
47. Kuiri, M.; Chakraborty, B.; Paul, A.; Das, S.; Sood, A.K.; Das, A. Enhancing photoresponsivity using MoTe$_2$-graphene vertical heterostructures. *Appl. Phys. Lett.* **2016**, *108*, 063506. [CrossRef]
48. Yin, L.; Zhan, X.; Xu, K.; Wang, F.; Wang, Z.; Huang, Y.; Wang, Q.; Jiang, C.; He, J. Ultrahigh sensitive MoTe$_2$ phototransistors driven by carrier tunneling. *Appl. Phys. Lett.* **2016**, *108*, 043503. [CrossRef]
49. Pradhan, N.R.; Rhodes, D.; Feng, S.; Xin, Y.; Memaran, S.; Moon, B.-H.; Terrones, H.; Terrones, M.; Balicas, L. Field-Effect Transistors Based on Few-Layered alpha-MoTe$_2$. *Acs Nano* **2014**, *8*, 5911–5920. [CrossRef]
50. Lin, Y.-F.; Xu, Y.; Lin, C.-Y.; Suen, Y.-W.; Yamamoto, M.; Nakaharai, S.; Ueno, K.; Tsukagoshi, K. Origin of Noise in Layered MoTe$_2$ Transistors and its Possible Use for Environmental Sensors. *Adv. Mater.* **2015**, *27*, 6612. [CrossRef]

51. Suk, J.W.; Kitt, A.; Magnuson, C.W.; Hao, Y.; Ahmed, S.; An, J.; Swan, A.K.; Goldberg, B.B.; Ruoff, R.S. Transfer of CVD-Grown Monolayer Graphene onto Arbitrary Substrates. *ACS Nano* **2011**, *5*, 6916–6924. [CrossRef] [PubMed]
52. Lou, J.; Xu, X.; Huang, Y.; Yu, Y.; Wang, J.; Fang, G.; Liang, J.; Fan, C.; Chang, C. Optically Controlled Ultrafast Terahertz Metadevices with Ultralow Pump Threshold. *Small* **2021**, *17*, e2104275. [CrossRef] [PubMed]
53. Cong, L.Q.; Singh, R. Spatiotemporal Dielectric Metasurfaces for Unidirectional Propagation and Reconfigurable Steering of Terahertz Beams. *Adv. Mater.* **2020**, *32*, 2001418. [CrossRef] [PubMed]
54. Manjappa, M.; Solanki, A.; Kumar, A.; Sum, T.C.; Singh, R. Solution-Processed Lead Iodide for Ultrafast All-Optical Switching of Terahertz Photonic Devices. *Adv. Mater.* **2019**, *31*, 1901455. [CrossRef] [PubMed]
55. Hu, Y.; Jiang, T.; Zhou, J.; Hao, H.; Sun, H.; Ouyang, H.; Tong, M.; Tang, Y.; Li, H.; You, J.; et al. Ultrafast terahertz transmission/group delay switching in photoactive WSe2-functionalized metaphotonic devices. *Nano Energy* **2020**, *68*, 104280. [CrossRef]
56. Krauss, T.F. Why do we need slow light? *Nat. Photonics* **2008**, *2*, 448–450. [CrossRef]
57. He, W.; Tong, M.; Xu, Z.; Hu, Y.; Cheng, X.A.; Jiang, T. Ultrafast all-optical terahertz modulation based on an inverse-designed metasurface. *Photonics Res.* **2021**, *9*, 1099–1108. [CrossRef]
58. Firstov, S.V.; Khegai, A.M.; Kharakhordin, A.V.; Alyshev, S.V.; Firstova, E.G.; Ososkov, Y.J.; Melkumov, M.A.; Iskhakova, L.D.; Evlampieva, E.B.; Lobanov, A.S.; et al. Compact and efficient O-band bismuth-doped phosphosilicate fiber amplifier for fiber-optic communications. *Sci. Rep.* **2020**, *10*, 11347. [CrossRef]
59. Liu, G.; Chang, C.; Qiao, Z.; Wu, K.; Zhu, Z.; Cui, G.; Peng, W.; Tang, Y.; Li, J.; Fan, C. Myelin Sheath as a Dielectric Waveguide for Signal Propagation in the Mid-Infrared to Terahertz Spectral Range. *Adv. Funct. Mater.* **2019**, *29*, 1807862. [CrossRef]
60. Zhang, J.; He, Y.; Liang, S.; Liao, X.; Li, T.; Qiao, Z.; Chang, C.; Jia, H.; Chen, X. Non-invasive, opsin-free mid-infrared modulation activates cortical neurons and accelerates associative learning. *Nat. Commun.* **2021**, *12*, 2730. [CrossRef]

**Disclaimer/Publisher's Note:** The statements, opinions and data contained in all publications are solely those of the individual author(s) and contributor(s) and not of MDPI and/or the editor(s). MDPI and/or the editor(s) disclaim responsibility for any injury to people or property resulting from any ideas, methods, instructions or products referred to in the content.

*Article*

# High-Speed THz Time-of-Flight Imaging with Reflective Optics

Hoseong Yoo [1], Jangsun Kim [2] and Yeong Hwan Ahn [1,*]

[1] Department of Physics and Department of Energy Systems Research, Ajou University, Suwon 16499, Republic of Korea
[2] Panoptics Corp., Seongnam 13516, Republic of Korea
* Correspondence: ahny@ajou.ac.kr

**Abstract:** In this study, we develop a 3D THz time-of-flight (TOF) imaging technique by using reflective optics to preserve the high-frequency components from a THz antenna. We use an Fe:InGaAs/InAlAs emitter containing relatively high-frequency components. THz-TOF imaging with asynchronous optical sampling (ASOPS) enables the rapid scanning of 100 Hz/scan with a time delay span of 100 ps. We characterize the transverse resolution using knife edge tests for a focal length of 5; the Rayleigh resolution has been measured at 1.0 mm at the focal plane. Conversely, the longitudinal resolution is determined by the temporal pulse width, confirmed with various gap structures enclosed by a quartz substrate. The phase analysis reveals that reflected waves from the top interface exhibit a phase shift when the gap is filled by high-indexed materials such as water but shows in-phase behavior when it is filled with air and low-indexed material. Our imaging tool was effective for inspecting the packaged chip with high lateral and longitudinal resolution. Importantly, the phase information in 2D and 3D images is shown to be a powerful tool in identifying the defect—in particular, delamination in the chip—which tends to be detrimental to the packaged chip's stability.

**Keywords:** terahertz imaging; nondestructive testing; time-of-flight

## 1. Introduction

Terahertz (THz) spectroscopy has emerged as a promising tool in a wide range of applications, such as material characterization, safety inspection, biomedical diagnosis, device inspection, telecommunication, and sensors [1–8]. In particular, the THz technique is desirable for the nondestructive testing (NDT) of the internal structures of objects. This is possible because THz waves are frequently transparent to the enclosures when they are made of nonconductive materials [9–20]. The recent technological advances in novel THz sources and detectors have increased the feasibility of effective THz imaging systems for practical application—for example, to identify ithe size, shape, and location of conductive objects and defects in insulating materials [21–29]. In addition, THz imaging, which is capable of 3D mapping on objects, has attracted particular interest because it allows users to locate their vertical as well as lateral positions. Coherent THz tomography, based on frequency-modulated CW (FMCW) methods, has been widely investigated using tunable or ensemble sources. However, it is limited in terms of longitudinal resolution [30–32]. In contrast, the time-of-flight methods based on time-domain spectroscopy (TDS) deliver better resolution in general, without using frequency sweeping methods. Here, the longitudinal resolution is determined by the temporal width of the pulsed laser source, enabling us to readily achieve 150 μm when the pulse width reaches 1 ps. Conversely, the relatively slow measurement time has limited the use of THz-TOF imaging for practical applications.

Recently, rapid THz-TOF imaging methods have been introduced for the inspection of packaged integrated circuits and defects in insulating materials [33,34]. An imaging rate of 100–200 Hz/pixels has been achieved based on the THz–TDS technique where fast time-delay scanning methods have been implemented. The optical sampling by cavity

tuning (OSCAT) system delivers a rapid time delay with a full range of 50 ps, whereby a single pulsed fiber source is split into two pulses used for the THz antenna and receiver, respectively [35,36]. Conversely, THz–TDS systems with asynchronous optical sampling (ASOPS) use two synchronized fiber lasers and deliver a time delay reaching 10 ns, which is determined by the laser repetition rate of 100 MHz [37,38]. The acquisition rate is determined by the repetition rate difference of the two fiber lasers that are selected to be approximately 100–500 Hz, depending on the signal-to-noise ratio. THz spectroscopy and imaging with the electronically controlled optical sampling (ECOPS) technique has been introduced using two independent lasers; however, fast imaging with galvanometer scanning has not been demonstrated yet [39–41].

For fast imaging with a speed of more than 100 Hz for each A-scan, it is important to incorporate galvano scanning methods in front of the scan lens. Recently, a THz antenna with a frequency range larger than 1 THz based on iron-doped InGaAs/InAlAs materials has become available [42]. Conventional THz imaging systems based on CW and pulsed sources have been implemented with refractive optics (transmission lenses). They were constructed of materials that were transparent against THz waves—for example, high resistivity float zone silicon (HRFZ-Si), quartz, fused silica (FS), and polymers [43]. However, there is noticeable attenuation in the high-frequency range larger than 1 THz. In addition, there is an inevitable reflection loss that is determined by the refractive index of the materials. Therefore, to optimize the new THz emitters both in terms of resolution and transmission efficiency, it is essential to adopt reflective optics instead of refractive optics, which will enable virtually loss-free configuration, especially in the high-frequency range. Conversely, the use of reflective optics in fast THz imaging systems has not yet been demonstrated.

In this study, we developed rapid THz-TOF imaging results consisting of reflective optics, in which a novel THz antenna delivers a better spatial resolution and enhanced signal-to-noise ratio. Using a focused lens of 50 mm, we characterized the imaging system by measuring the resolution in the transverse and longitudinal directions. We also introduced a novel phase mapping technique that is efficient in identifying delamination defects in the packaged semiconducting chips.

## 2. Experimental Setup

A schematic of the THz-TOF imaging equipment is illustrated in Figure 1a. For the rapid THz–TDS system, we used a commercialized ASOPS instrument (TERA-ASOPS, MenloSystems GmbH), consisting of two-independent femtosecond fiber lasers with a wavelength of 1560 nm, operating with a repetition rate of 100 MHz. The ASOPS system allows a full time-delay range of 10 ns with a rate of 100–500 Hz/pixel, which we fixed at 100 Hz/pixel throughout the experiments. We recorded the data with the time span of 100 ps (out of the full delay scan range of 10 ns), as shown in Figure 1b, because the transverse resolution is limited by the depth of focus, as will be discussed later. THz-TOF with ASOPS has a significant advantage in terms of large-depth profiling, whereby it remains restricted by the focal depth of the scan imaging system. In particular, we incorporated a novel THz antenna with a large bandwidth based on the iron-doped InGaAs/InAlAs material, which delivers a large amount of high-frequency components, as shown in Figure 1c. We note that there is a hemispherical lens attached to the commercialized antenna (for both emitter and detector), which is used to guide the THz beams with a divergence angle of 29°. High-frequency components are critical for obtaining images with high resolutions. In general, conventional refractive optics in the image scanning system are typically made of HRFZ-Si or polymethylpentene (TPX) and attenuate a large portion of the high-frequency components. In the TOF imaging system, the THz pulse passes through the lenses four times, including transmission through a scan lens (back and forth) in front of the sample and through two lenses in front of the emitter and detector.

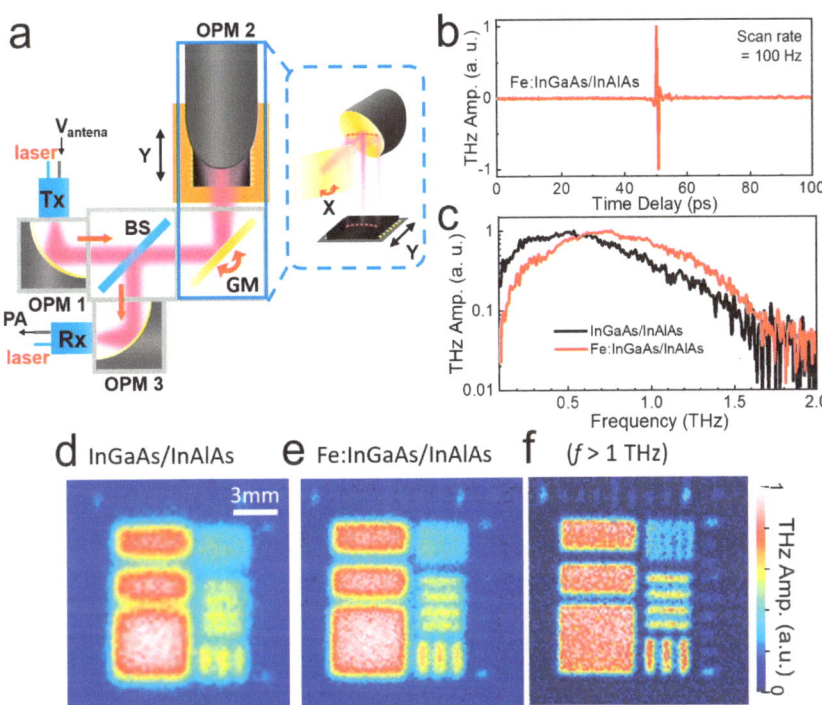

**Figure 1.** (**a**) Schematic illustration of the THz-TOF imaging setup with all-reflective optics. (**b**) A representative THz reflection amplitude as a function of time delay (A-scan). (**c**) THz reflection spectra for both antenna types (**c**–**e**) Time-integrated reflection amplitude images on a test pattern (C-scan) with old antenna (**d**), new antenna (**e**), and high-frequency filtered image with the new antenna (**f**). GM: galvanometer, BS: beamsplitter, OPM: off-axis parabolic mirror, PA: current preamplifier.

Here, we developed THz imaging systems based on pure reflective optics (Figure 1a), except for the hemispherical lens attached to the commercialized antenna. We used three off-axis parabolic mirrors (OPMs) and a one-axis galvano scanner. An OPM with a diameter of 25 mm and a focal length of 50 mm was used in front of the THz emitter (denoted by Tx) to collimate the THz beams from the emitter. The other OPM, with a diameter of 50 mm and a focal length of 50 mm, was used to focus the THz beam onto the samples. To simplify the imaging systems while obtaining a speed that is limited by the ASOPS THz–TDS system, we combined fast galvano scanning for the $x$-axis and slow stage scanning for the $y$-axis. The reflected THz signal was collected using a 50:50 beam splitter (Tydex Inc. St. Peterburg, Russia), which was focused onto the THz receiving antenna (denoted by Rx) by using another OPA. The same specification used for the emitter was used for the detector. We note that distortion in the raster-scanned image occurs particularly when the scan position is far from the center. This is because conventional OPMs are designed to focus the collimated beams, not to raster-scan with the galvano scanning system. Correction of the image distortion originating from the imperfect $f$-$\theta$ scan lens and galvano scanning systems has been widely investigated [44]. It is relatively simple in our case, because the correction is needed along one direction, because we combined the sample stage scanning for another axis ($y$-axis). The image distortion differs, depending on the scan direction. In this case, we chose to scan along the horizontal line (inset, Figure 1a). Distortion in the image can be corrected post-measurement or in situ by slightly adjusting the $y$-position of the sample stage along the curved line. In addition, the signal amplitudes were normalized by reference signals—in other words, by the position-dependent reflection signal from the gold plate.

The current signal from the antenna was amplified using a fast current preamplifier (DLPCA-100; FEMTO Messtechnik GmbH) and digitized with a data acquisition card (ATS9462, Alazar Tech. Inc., Pointe-Claire, QC, Canada) with a rate of 10 MHz. This allowed us to obtain a time-delay step of 0.1 ps. The collected data contained phase-sensitive THz amplitudes as a function of the 3D parameters, such as the time delay, $x$-axis, and $y$-axis. We applied the point average to improve the signal-to-noise ratio, leading to a measurement rate of 5–10 Hz/pixel, while the system delivered 100 Hz/pixel. We used a delay-scan range of $\Delta T = 100$ ps, which corresponds to a vertical range of 15 mm in the case of free-space propagation. Here, the time delay can be converted into depth information, in which 1 ps corresponds to $0.15/n$ mm in the reflection geometry, where $n$ is the refractive index of the materials [33,34]. The images were recorded as a single binary file, which can be reconstructed using home-built analyzing software. The phase-sensitive THz amplitudes can be converted into the envelope signal by using the Hilbert transformation—that is, from the complex THz signal $\tilde{E}_{THz} = \text{Re}(\tilde{E}_{THz}) + iH[\text{Re}(\tilde{E}_{THz})]$, where $H$ is the Hilbert transformation [45].

Shown in Figure 1d,e are C-scan images for time-integrated reflection amplitudes on the test alignment pattern (Au pattern on Si substrate), when we used old and new antennae, respectively. Clearly, the spatial resolution improved with the new antenna, having more high-frequency components. The resolution improves even further if we filter the high-frequency component (>1 THz), as shown in Figure 1f. Obviously, it is essential for practical applications to develop THz sources and detectors having higher-frequency components while delivering fast scan rates. In addition, the use of reflective optics allows us to reduce the reflection losses originating from the finite refractive index of the lens material.

## 3. Results

### 3.1. Measurements of Transverse Resolution

We first characterized the transverse resolution of our imaging system by comparing it with the old emitter. As shown in Figure 2a, we used a metal plate with a knife edge to measure the transverse resolution of the system. We obtained THz images for different heights ($z$) of the metallic plate, from $z = -25$ to $z = +25$ mm, where $z = 0$ indicates the location of the focus plane, which was positioned 5 cm below the lens. Figure 2b illustrates the line profile of the reflected amplitude signal as a function of the lateral position ($x$), extracted from the spectrally integrated C-scan images shown in the inset. Here, the vertical location was fixed at $z = 0$ mm (Figure 2a). The reflection signal was high when it was reflected by the metal plane ($x < 4$ mm), whereas it was suppressed when the focused beam was positioned outside of the metal (on the right side of the knife edge).

The line profile was fitted with the error function to obtain the transverse resolution of the focused THz beams. We determined the transverse Rayleigh resolution ($R_T$) using the knife-edge 10–90% transition in THz amplitude in Figure 2b. By fitting the curve, the $R_T$ of the reflected THz amplitude was measured at 1.0 mm, which corresponds to the Rayleigh half-pitch resolution of 0.5 mm [46,47]. The extracted $R_T$ as a function of the vertical position ($z$) of the metal plane is plotted for both antenna types (Figure 2c). The minimum resolution was achieved at approximately $z = 0$ mm and it increased when the metal plane was away from the focused plane. For example, it increased up to $R_T = 2.3$ mm for $z = -20$ mm. In contrast, we could readily achieve $R_T$ in the range of 1.0–1.1 mm within a depth range of 10 mm (from $z = -5$ mm to 5 mm). We also note that the resolution increased noticeably by introducing the novel antenna (red circles) when compared to the spatial resolution of the old antenna (black squares); in other words, the transverse resolution improved by approximately 33% compared to that of the old antenna having $R_T = 1.5$ mm.

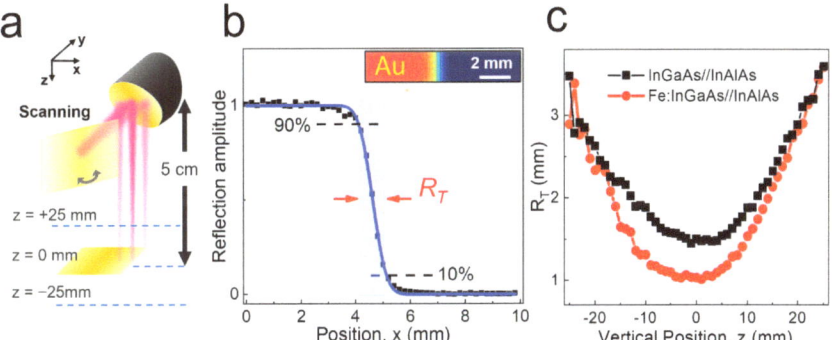

**Figure 2.** (a) Schematic illustration for transverse resolution measurement for different z-locations with a knife edge. (b) THz signal (black squares) as a function of position (x) when the metal plate is located at z = 0 mm. The blue solid line is the fitted curve with the error function. $R_T$ denotes the transverse resolution defined with 10–90% transition of THz amplitudes. (inset) Time-integrated C-scan image of the reflected wave envelope from the knife edge. Au indicates gold surface. (c) The transverse resolution ($R_T$) as a function of the vertical position (z) of the metal plate, extracted from the fitting result in (c) for both InGaAs/InAlAs and Fe:InGaAs/InAlAs antennas.

Obviously, the spatial resolution will improve if we choose the high-frequency components, as shown in Figure 1f. Conversely, the TOF information is no longer available upon the spectral analysis. Most of the literature on 3D imaging has focused on the resolution in terms of the specific frequency range [31,32,48]; a direct comparison to our results is restricted here. Focusing on the TOF imaging systems, the focal spot size of 2.3 mm has been reported with the refractive lens systems [16]. In our work, we optimized the resolution in the THz-TOF systems based on the novel THz antenna. This was possible by adopting the reflective optics, which enables us to preserve the high-frequency components, particularly without sacrificing the measurement speed with the help of the galvano scanning methods.

*3.2. Measurement of Longitudinal Resolution*

The longitudinal resolution is determined by the temporal width of the THz pulse, as shown in Figure 3, whereas the transverse resolution is determined by the imaging optics and wavelength of the THz (Figure 2). To address the longitudinal resolution, we fabricated a gap structure with various sizes, in which the gap is filled by air, enclosed by two quartz plates. Here, we varied the air gap ($d_{gap}$) from 0 µm to 320 µm by tilting the top plate with respect to the bottom plate. The A-scan data for the reflected THz amplitude with $d_{gap}$ = 305 µm and 116 µm are demonstrated in Figure 3a,b, respectively. The THz envelope extracted via the Hilbert transformation is characterized by a double peak for $d_{gap}$ = 305 µm, with the peaks well separated from each other.

By fitting with the Gaussian function, the difference between the two peaks is found at $\Delta T$ = 2.03 ps, which is consistent with the measured $d_{gap}$, whereas the full width at half maximum of the respective envelope is found at 1.0 ps. The envelope function was fitted well with the Gaussian function with a standard deviation of less than 1%. Conversely, a careful fitting process is required when $d_{gap}$ is close to the longitudinal resolution of the system, in which we found $\Delta T$ = 0.77 ps for $d_{gap}$ = 116 µm (Figure 3b). The fitted gap size ($d_{fit}$) as a function of $d_{gap}$ is plotted in Figure 3c, in which the fitted value matches well with $d_{gap}$. Conversely, when $d_{gap} < w$, where $w$ is the $1/e^2$ halfwidth of a single Gaussian function, the gap size information is no longer available from the fitting process with a double Gaussian function. Instead, it can be estimated from the broadening of the reflected pulse by obtaining $w$ with a single Gaussian function (inset, Figure 3c). It can be seen that $w$ increased from $w$ = 116 µm of $d_{gap}$ = 8.7 µm to $w$ = 170 µm of $d_{gap}$ = 116 µm. Consequently, the linear relationship between $w$ and $d_{gap}$ helps us to obtain the gap size information down to a near-zero gap size.

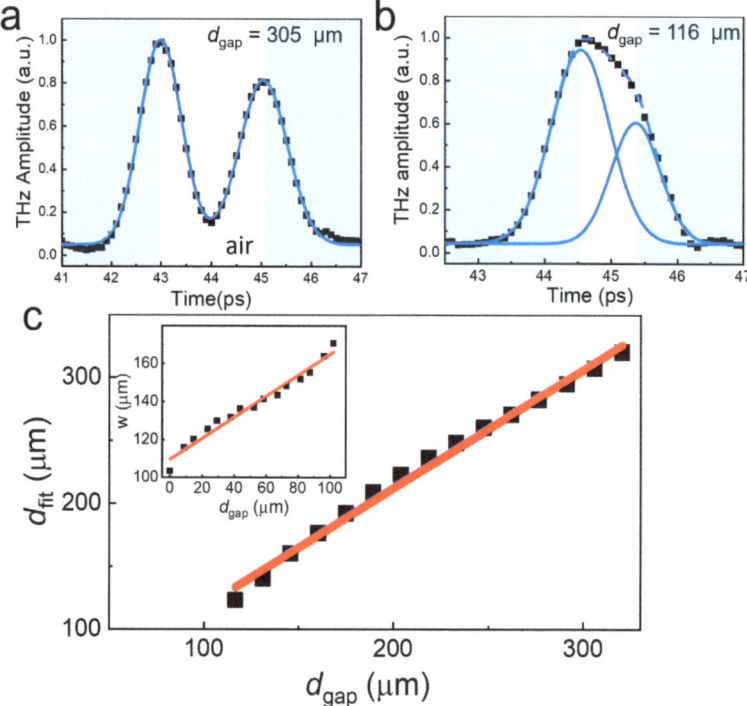

**Figure 3.** (**a**) A-scan envelope of the reflected waves for the device with an air gap ($d_{gap}$) = 305 μm enclosed by two quartz substrates. The blue line is a fit to the data with a double Gaussian function. (**b**) A-scan envelope of the reflected waves with $d_{gap}$ = 116 μm. (**c**) The estimated gap width ($d_{fit}$) by fitting the THz reflection envelope with the double Gaussian functions as a function of $d_{gap}$ as shown in (**a**,**b**) The red line is a fit to the data with a linear function. (inset). The $1/e^2$ halfwidth ($w$) of the Gaussian function was obtained by fitting the THz envelope with a single Gaussian function.

### 3.3. Phase of Reflected Pulses from Gap Structures with Variable Height

Importantly, the phase information obtained in nondestructive testing with the THz–TDS system allows us to address the nature of the gap enclosed by the insulating materials—that is, whether it is filled by the air or by the polymer, whose refractive index is larger than that of the enclosure. For example, it is critically important to address the types of defects in the packaged semiconducting chips, namely whether the gap is void because of delamination or it is filled by epoxy. It is very important to address this issue because reliable mechanical support for the packaged chip is essential for the physical protection of the device, distribution of electrical power, and heat dissipation of circuits [41]. This is possible with the THz-TOF imaging because the waves experience a 180° phase shift when they are reflected against materials with a relatively higher dielectric constant.

First, we show the depth profiling directly from the B-scan images for the samples consisting of two quartz substrates with an air gap in the middle (Figure 4a). The THz signal was plotted as a function of position $x$ and time delay ($T$). Here, we fitted the envelope function in the time domain with a Gaussian function and plotted the peak values when they were above a threshold value for each pixel. Therefore, the position in the $T$ axis denotes the peak position (which has been broadened to be three points for clarity), whereas the color scale denotes the reflection amplitude at the peak position. The THz reflection was dominated by the signal from the air–quartz and quartz–air interfaces and so there is a gap structure between the two substrates.

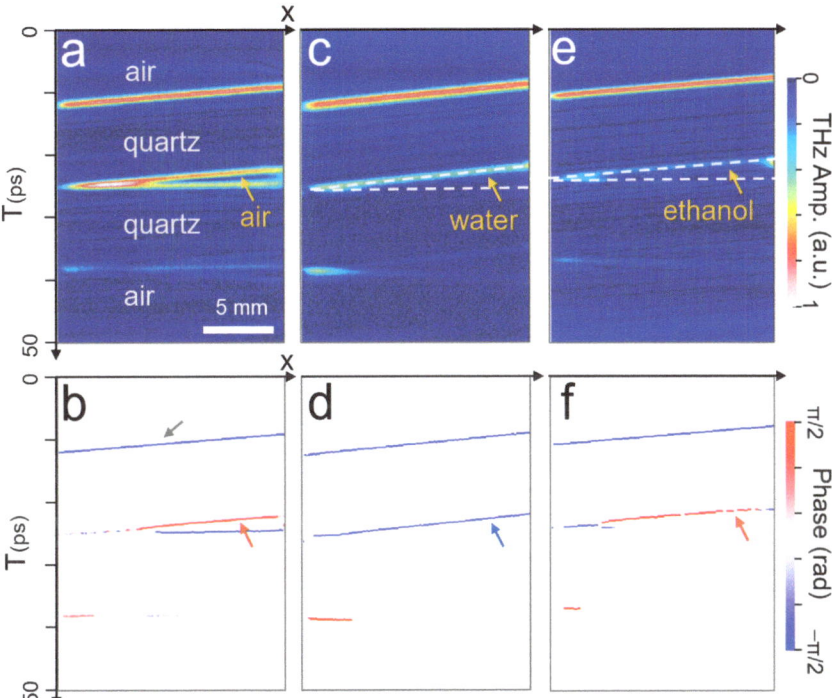

**Figure 4.** (**a**,**b**) B-scan images for the reflected amplitude (**a**) and phase (**b**) for the samples with a tilted gap that is filled with air. (**c**,**d**) B-scan images for amplitude (**c**) and phase (**d**) when the gap is filled by water solution. (**e**,**f**) B-scan images for amplitude (**e**) and phase (**f**) when the gap is filled with ethanol whose index is lower than that of quartz.

In Figure 4b, we show the B-scan image in terms of the phase function obtained by the Hilbert transformation. Here, the positive polarity (in-phase) indicates the reflection from the surfaces with the lower-indexed medium, whereas the negative polarity (out-of-phase) implies the reflection from the higher-indexed medium. For example, the reflection at the top surface of the quartz plate (indicated by a grey arrow) shows the negative phase because the index of the quartz is higher than that of the air. Importantly, the positive phase in the upper interface of the gap region (red arrow in Figure 4b) indicates that the gap is filled by dielectric material with the lower-indexed medium, which, in this case, is the air.

In contrast, the phase of the upper interface of the gap exhibits a negative value (Figure 4c,d), indicated by a blue arrow in (d), when the gap is filled by water ($n_{water}$~2.2). This is because the index of the water is higher than that of the quartz ($n_{quartz}$~1.9). We also note that the signal from the bottom interface of the gap is significantly suppressed because of the large attenuation by the water layer. We note that this is similar to the case when the gap is filled by a polymer such as epoxy which is highly absorptive in the THz frequency range [33]. Finally, we show the B-scan images for amplitude and phase (Figure 4e,f) when the gap is filled by ethanol ($n_{ethanol}$~1.6), whose index is slightly smaller than that of quartz [49]. Now, the phase at the top interface of the gap shows a positive value (red arrow), as in the case of the air gap. We also note that the amplitude of the reflected waves (Figure 4e) is small compared to that of the air gap because the dielectric contrast between the quartz and ethanol is relatively low.

## 3.4. TOF Imaging of Representative Packaged Chips

Figure 5 provides an example of THz-TOF imaging for a packaged chip with a relatively large thickness. We note that the ASOPS system, with a large scan range, enables NDT on objects with a large thickness. In Figure 5a, a photograph of a semiconductor package with dimensions of 8 mm × 35 mm × 3.3 mm is shown. Figure 5b shows a time-integrated C-scan image for the device, and Figure 5c is the corresponding B-scan image (as a function of $y$ and $T$) along the $y$-axis at the center of the device. Figure 5d shows a series of C-scan images for different $z$ locations for the sample—that is, the THz reflection images at the top, chip, and leadframe surfaces appear at $T$ = 14 ps, 31 ps, and 34 ps, respectively. Considering the refractive index of the packaging material ($n$ = 2.0), the TOF time differences reveal that the chip surface is located at 1.28 mm ($\Delta T$ = 17 ps) beneath the package top surface. Conversely, the leadframe surface appears at 1.5 mm ($\Delta T$ = 20 ps) below the top surface. Consequently, the THz-TOF data can then be reconstructed into 3D images (Figure 5e). Here, different colors were assigned for the reflection amplitude, with red and white colors indicating the high reflection amplitude. The chip location is clearly identified by the very large reflection amplitude, which is because semiconducting chips have a larger reflectance than insulating materials used as enclosers. In this regard, 3D mapping techniques, wrapped with their reflection amplitude, will be very effective in locating various structures embedded in the insulating materials, potentially including their material information.

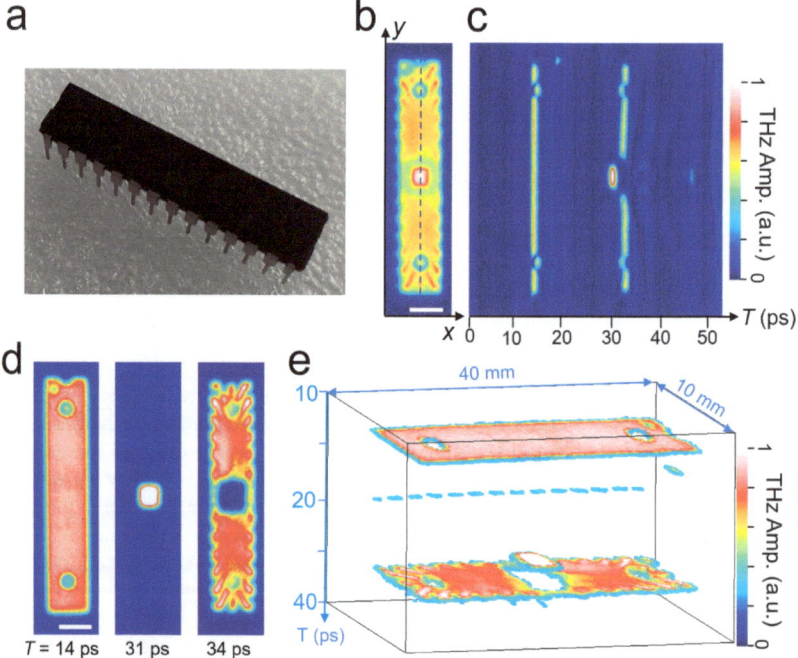

**Figure 5.** (**a**) Photograph of the chip with dimensions of 8 mm × 35 mm × 3.3 mm. (**b**) Time-integrated C-scan envelopes for the packaged device. (**c**) B-scan images along the dashed line in (**b**). Scale bar = 5 mm. (**d**) A series of C-scan ($xy$) images for T = 14, 31, and 34 ps, respectively. (**e**) Reconstructed 3D image for the device.

## 3.5. Peak Envelopes Modulated with Phase Polarity for Defect Identification

An example of identifying defects in packaged chips is illustrated in Figure 6, with a packaged device with the dimensions 20 mm × 20 mm × 1 mm. For improved image quality, we used a sample stage scan for both the $x$- and $y$-axis because the chip size is rather large. Figure 6a illustrates the time-integrated C-scan image of the device, which contains a region for the chip in the middle that is represented by a bright reflection region. In many cases, the presence of defects is not clear from the C-scan images because most of the packaged chips show similar irregular patterns outside the chip area, regardless of the presence of defects. Conversely, B-scan (x-T plot) images clearly exhibit the presence of gap structures in the middle of the packages (Figure 6b). Here, the THz amplitude is plotted along the dashed line in Figure 6a and as a function of time delay. Two different types of gap regions are clearly identified (orange and grey arrows), with gap sizes corresponding to $\Delta T = 1.9$ ps and 1.7 ps, respectively. As previously discussed, the types of defects can be addressed by phase mapping (Figure 6c). In other words, it shows that the phase at the top interface exhibited a positive value (in phase) in the region denoted by an orange arrow, whereas it shows a negative value (out of phase) in the region denoted by a grey arrow. Identifying the types of defects in the packaged chips—that is, whether the gap is a void or epoxy-rich region—is our primary concern. In particular, locating the delaminated area is essential because it could cause detrimental failures to the chips in terms of mechanical stability and heat dissipation [41]. This is possible because THz waves suffer from a phase change when they are reflected at an interface that has a large dielectric constant.

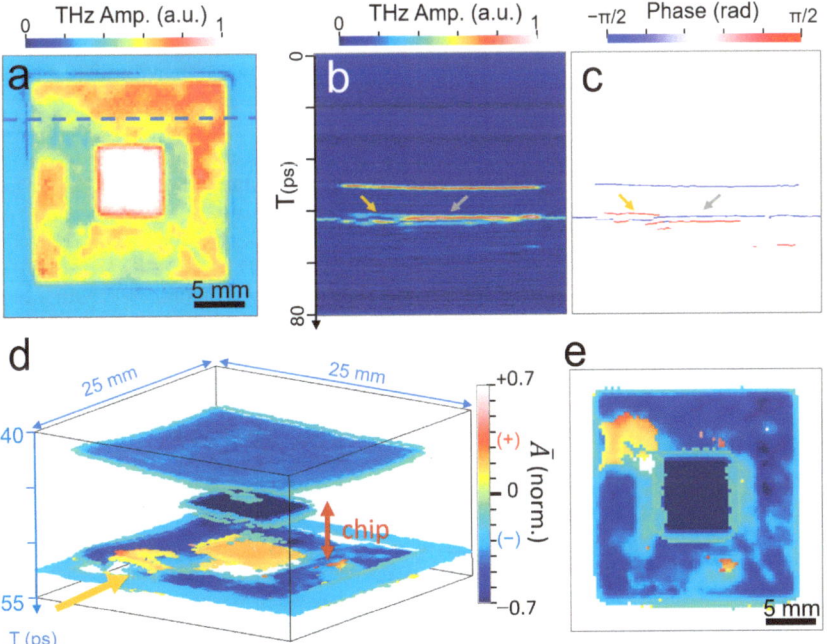

**Figure 6.** (a) Time-integrated C-scan image for a packaged chip with dimensions of 20 mm × 20 mm × 1 mm. (b) B-scan image for the reflected amplitude taken along the dashed line in (a). (c) B-scan image for the phase of the reflected waves. The orange and grey arrows indicate the void and epoxy-rich regions, respectively. (d) 3D image of reflected amplitudes in terms of the peak envelope modulated with phase polarity. (e) C-scan image for the peak envelope with phase polarity for the second peaks appearing in the time domain to highlight the delamination area (red) exclusively.

The image was reconstructed as a 3D image by plotting the peak envelope modulated with phase polarity ($\overline{A}$) as a function of $x$, $y$, and $T$ (Figure 6d). In other words, we plotted $\overline{A} = A_{THz} \times sign(\phi)$, where $A_{THz}$ is the normalized peak envelope and $sign(\phi)$ is the sign of the phase at the peak. Here, $A_{THz}$ was normalized by that of the reflection at the top surface of the chip, which has the highest reflection value. The positive (in-phase) and negative (out-of-phase) polarities again appeared, depending on the types of reflection.

For example, the top surface of the package and that of the embedded chip (colored in dark blue at the center) exhibit negative polarity, whereas positive polarity is shown at the bottom of the chip (colored in orange at the center). This is consistent with the result shown in Figure 6c. More importantly, a positive phase appears in the region indicated by an arrow (orange), which clearly indicates the presence of delamination, whereas the epoxy-rich regions were not clearly identified in the 3D image. This is because the phase for the epoxy-rich region is the same as that of defect-free areas. Figure 6e shows a 2D image (C-scan), in which we plot the phase-modulated envelope for the second peaks appearing along the $T$ axis. This is to exclude reflection from the top package surface and focus on the interface between the top and bottom plastic enclosures. This is the layer in which various types of defects occur during the package sealing process. As a result, both 2D and 3D imaging, based on phase-modulated peak envelopes, are shown to be unique and powerful tools for identifying the types of defects and, in particular, delamination, which is critical to chip stability.

## 4. Conclusions

We developed rapid THz-TOF imaging results consisting of reflective optics, in which a novel THz antenna based on Fe:InGaAs/InAlAs delivers a better spatial resolution and enhanced signal-to-noise ratio. We also incorporated the ASOPS technique, enabling a rapid A-scan rate of 100 Hz/scan with a large time-delay span of 100 ps. We characterized the transverse resolution with a focal length of 5 cm; the resolution in terms of the knife-edge 10–90% transition reaches 1.0 mm at the focal plane. In contrast, the longitudinal resolution reached 100 μm, which is determined by the temporal pulse width, whereas the gap width can be estimated indirectly by measuring the broadening of the pulse when the gap size is lower than the pulse width. We studied the phase shift in the reflected THz waves by using the gap structure when it is filled with air, water, and ethanol solution. The novel phase analysis revealed that we can identify the types of reflection depending on the dielectric contrast at the interface. Our 3D imaging tool was very effective in the nondestructive inspection of a packaged semiconducting chip with a high transverse and longitudinal resolution. Importantly, the phase information in 2D and 3D images is demonstrated to be powerful for identifying the delaminated region, which has a large phase contrast with opposite polarity.

**Author Contributions:** Conceptualization, Y.H.A.; Data curation, H.Y. and Y.H.A.; Formal analysis, H.Y. and Y.H.A.; Investigation, H.Y. and Y.H.A.; Methodology, H.Y.; Resources, J.K.; Software, Y.H.A.; Writing—original draft, H.Y., J.K. and Y.H.A.; Writing—review and editing, H.Y., J.K. and Y.H.A. All authors have read and agreed to the published version of the manuscript.

**Funding:** This work was supported by the Midcareer Researcher Program (2020R1A2C1005735) and Basic Science Research Program (2021R1A6A1A10044950) through a National Research Foundation grant funded by the Korean Government. It is also supported by the GRRC Program (GRRCA-JOU2022B01, Photonics-Medical Convergence Technology Research Center) of Gyeonggi Province, Republic of Korea.

**Institutional Review Board Statement:** Not applicable.

**Informed Consent Statement:** Not applicable.

**Data Availability Statement:** Not applicable.

**Conflicts of Interest:** The authors declare no conflict of interest.

## References

1. Ogawa, Y.; Hayashi, S.; Oikawa, M.; Otani, C.; Kawase, K. Interference terahertz label-free imaging for protein detection on a membrane. *Opt. Express* **2008**, *16*, 22083–22089. [CrossRef] [PubMed]
2. Zhong, S.; Shen, Y.C.; Ho, L.; May, R.K.; Zeitler, J.A.; Evans, M.; Taday, P.F.; Pepper, M.; Rades, T.; Gordon, K.C.; et al. Non-destructive quantification of pharmaceutical tablet coatings using terahertz pulsed imaging and optical coherence tomography. *Opt. Lasers Eng.* **2011**, *49*, 361–365. [CrossRef]

3. Tomaino, J.L.; Jameson, A.D.; Paul, M.J.; Kevek, J.W.; Van Der Zande, A.M.; Barton, R.A.; Choi, H.; McEuen, P.L.; Minot, E.D.; Lee, Y.S. High-contrast imaging of graphene via time-domain terahertz spectroscopy. *J. Infrared Millim. Terahertz Waves* **2012**, *33*, 839–845. [CrossRef]
4. Park, S.J.; Hong, J.T.; Choi, S.J.; Kim, H.S.; Park, W.K.; Han, S.T.; Park, J.Y.; Lee, S.; Kim, D.S.; Ahn, Y.H. Detection of microorganisms using terahertz metamaterials. *Sci. Rep.* **2014**, *4*, 1–7. [CrossRef] [PubMed]
5. Park, S.H.; Jang, J.W.; Kim, H.S. Non-destructive evaluation of the hidden voids in integrated circuit packages using terahertz time-domain spectroscopy. *J. Micromechanics Microeng.* **2015**, *25*, 095007. [CrossRef]
6. Kim, H.S.; Ha, N.Y.; Park, J.Y.; Lee, S.; Kim, D.S.; Ahn, Y.H. Phonon-Polaritons in Lead Halide Perovskite Film Hybridized with THz Metamaterials. *Nano Lett.* **2020**, *20*, 6690–6696. [CrossRef]
7. Wang, L. Terahertz imaging for breast cancer detection. *Sensors* **2021**, *21*, 6465. [CrossRef]
8. Jun, S.W.; Ahn, Y.H. Terahertz thermal curve analysis for label-free identification of pathogens. *Nat. Commun.* **2022**, *13*, 3470. [CrossRef]
9. Kawase, K.; Ogawa, Y.; Watanabe, Y.; Inoue, H. Non-destructive terahertz imaging of illicit drugs using spectral fingerprints. *Opt. Express* **2003**, *11*, 2549–2554. [CrossRef]
10. Federici, J.F.; Schulkin, B.; Huang, F.; Gary, D.; Barat, R.; Oliveira, F.; Zimdars, D. THz imaging and sensing for security applications—Explosives, weapons and drugs. *Semicond. Sci. Technol.* **2005**, *20*, S266–S280. [CrossRef]
11. Zhong, H.; Xu, J.; Xie, X.; Yuan, T.; Reightler, R.; Madaras, E.; Zhang, X.C. Nondestructive defect identification with terahertz time-of-flight tomography. *IEEE Sens. J.* **2005**, *5*, 203–207. [CrossRef]
12. Karpowicz, N.; Redo, A.; Zhong, H.; Li, X.; Xu, J.; Zhang, X.C. Continuous-wave terahertz imaging for non-destructive testing applications. In Proceedings of the Joint 30th International Conference on Infrared and Millimeter Waves and 13th International Conference on Terahertz Electronics, IRMMW-THz 2005, Williamsburg, VI, USA, 19–23 September 2005; pp. 329–330.
13. Shen, Y.C.; Lo, T.; Taday, P.F.; Cole, B.E.; Tribe, W.R.; Kemp, M.C. Detection and identification of explosives using terahertz pulsed spectroscopic imaging. *Appl. Phys. Lett.* **2005**, *86*, 241116. [CrossRef]
14. Schirmer, M.; Fujio, M.; Minami, M.; Miura, J.; Araki, T.; Yasui, T. Biomedical applications of a real-time terahertz color scanner. *Biomed. Opt. Express* **2010**, *1*, 354–366. [CrossRef] [PubMed]
15. Kawase, K.; Shibuya, T.; Hayashi, S.; Suizu, K. THz imaging techniques for nondestructive inspections. *Comptes Rendus Phys.* **2010**, *11*, 510–518. [CrossRef]
16. Jin, K.H.; Kim, Y.G.; Cho, S.H.; Ye, J.C.; Yee, D.S. High-speed terahertz reflection three-dimensional imaging for nondestructive evaluation. *Opt. Express* **2012**, *20*, 25432–25440. [CrossRef]
17. Fan, S.; Li, T.; Zhou, J.; Liu, X.; Liu, X.; Qi, H.; Mu, Z. Terahertz non-destructive imaging of cracks and cracking in structures of cement-based materials. *AIP Adv.* **2017**, *7*, 115202. [CrossRef]
18. Ahi, K.; Shahbazmohamadi, S.; Asadizanjani, N. Quality control and authentication of packaged integrated circuits using enhanced-spatial-resolution terahertz time-domain spectroscopy and imaging. *Opt. Lasers Eng.* **2018**, *104*, 274–284. [CrossRef]
19. Fuse, N.; Sugae, K. Non-destructive terahertz imaging of alkali products in coated steels with cathodic disbanding. *Prog. Org. Coat.* **2019**, *137*, 105334. [CrossRef]
20. Zhang, J.Y.; Ren, J.J.; Li, L.J.; Gu, J.; Zhang, D.D. THz imaging technique for nondestructive analysis of debonding defects in ceramic matrix composites based on multiple echoes and feature fusion. *Opt. Express* **2020**, *28*, 19901–19915. [CrossRef]
21. Karpowicz, N.; Zhong, H.; Zhang, C.; Lin, K.I.; Hwang, J.S.; Xu, J.; Zhang, X.C. Compact continuous-wave subterahertz system for inspection applications. *Appl. Phys. Lett.* **2005**, *86*, 054105. [CrossRef]
22. Karpowicz, N.; Zhong, H.; Xu, J.; Lin, K.I.; Hwang, J.S.; Zhang, X.C. Comparison between pulsed terahertz time-domain imaging and continuous wave terahertz imaging. *Semicond. Sci. Technol.* **2005**, *20*, S293–S299. [CrossRef]
23. Lien Nguyen, K.; Johns, M.L.; Gladden, L.F.; Worrall, C.H.; Alexander, P.; Beere, H.E.; Pepper, M.; Ritchie, D.A.; Alton, J.; Barbieri, S.; et al. Three-dimensional imaging with a terahertz quantum cascade laser. *Opt. Express* **2006**, *14*, 2123–2129. [CrossRef] [PubMed]
24. Kim, J.-Y.; Song, H.-J.; Yaita, M.; Hirata, A.; Ajito, K. CW-THz vector spectroscopy and imaging system based on 1.55-μm fiber-optics. *Opt. Express* **2014**, *22*, 1735–1741. [CrossRef] [PubMed]
25. Lee, I.S.; Lee, J.W. Nondestructive internal defect detection using a CW-THz imaging system in XLPE for power cable insulation. *Appl. Sci.* **2020**, *10*, 2055. [CrossRef]
26. Mathanker, S.K.; Weckler, P.R.; Wang, N. Terahertz (THz) applications in food and agriculture: A review. *Trans. ASABE* **2013**, *56*, 1213–1226.
27. Ok, G.; Park, K.; Kim, H.J.; Chun, H.S.; Choi, S.W. High-speed terahertz imaging toward food quality inspection. *Appl. Opt.* **2014**, *53*, 1406–1412. [CrossRef]
28. Wang, K.; Sun, D.W.; Pu, H. Emerging non-destructive terahertz spectroscopic imaging technique: Principle and applications in the agri-food industry. *Trends Food Sci. Technol.* **2017**, *67*, 93–105. [CrossRef]
29. Afsah-Hejri, L.; Hajeb, P.; Ara, P.; Ehsani, R.J. A Comprehensive Review on Food Applications of Terahertz Spectroscopy and Imaging. *Compr. Rev. Food Sci. Food Saf.* **2019**, *18*, 1563–1621. [CrossRef]
30. Nagatsuma, T.; Nishii, H.; Ikeo, T. Terahertz imaging based on optical coherence tomography [invited]. *Photonics Res.* **2014**, *2*, B64–B69. [CrossRef]
31. Cristofani, E.; Friederich, F.; Wohnsiedler, S.; Matheis, C.; Jonuscheit, J.; Vandewal, M.; Beigang, R. Nondestructive testing potential evaluation of a terahertz frequency-modulated continuous-wave imager for composite materials inspection. *Opt. Eng.* **2014**, *53*, 031211. [CrossRef]

32. Yahng, J.S.; Park, C.S.; Lee, H.D.; Kim, C.S.; Yee, D.S. High-speed frequency-domain terahertz coherence tomography. *Opt. Express* **2016**, *24*, 1053–1061. [CrossRef] [PubMed]
33. Yim, J.H.; Kim, S.Y.; Kim, Y.; Cho, S.; Kim, J.; Ahn, Y.H. Rapid 3d-imaging of semiconductor chips using thz time-of-flight technique. *Appl. Sci.* **2021**, *11*, 4770. [CrossRef]
34. Kim, H.S.; Baik, S.Y.; Lee, J.W.; Kim, J.; Ahn, Y.H. Nondestructive tomographic imaging of rust with rapid thz time-domain spectroscopy. *Appl. Sci.* **2021**, *11*, 10594. [CrossRef]
35. Hochrein, T.; Wilk, R.; Mei, M.; Holzwarth, R.; Krumbholz, N.; Koch, M. Optical sampling by laser cavity tuning. *Opt. Express* **2010**, *18*, 1613–1617. [CrossRef] [PubMed]
36. Wilk, R.; Hochrein, T.; Koch, M.; Mei, M.; Holzwarth, R. OSCAT: Novel technique for time-resolved experiments without moveable optical delay lines. *J. Infrared Millim. Terahertz Waves* **2011**, *32*, 596–602. [CrossRef]
37. Yasui, T.; Saneyoshi, E.; Araki, T. Asynchronous optical sampling terahertz time-domain spectroscopy for ultrahigh spectral resolution and rapid data acquisition. *Appl. Phys. Lett.* **2005**, *87*, 061101. [CrossRef]
38. Bartels, A.; Thoma, A.; Janke, C.; Dekorsy, T.; Dreyhaupt, A.; Winnerl, S.; Helm, M. High-resolution THz spectrometer with kHz scan rates. *Opt. Express* **2006**, *14*, 430–437. [CrossRef]
39. Kim, Y.; Yee, D.-S. High-speed terahertz time-domain spectroscopy based on electronically controlled optical sampling. *Opt. Lett.* **2010**, *35*, 3715–3717. [CrossRef]
40. Pałka, N.; Maciejewski, M.; Kamiński, K.; Piszczek, M.; Zagrajek, P.; Czerwińska, E.; Walczakowski, M.; Dragan, K.; Synaszko, P.; Świderski, W. Fast THz-TDS Reflection Imaging with ECOPS—Point-by-Point versus Line-by-Line Scanning. *Sensors* **2022**, *22*, 8813. [CrossRef]
41. Yahyapour, M.; Jahn, A.; Dutzi, K.; Puppe, T.; Leisching, P.; Schmauss, B.; Vieweg, N.; Deninger, A. Fastest Thickness Measurements with a Terahertz Time-Domain System Based on Electronically Controlled Optical Sampling. *Appl. Sci.* **2019**, *9*, 1283. [CrossRef]
42. Globisch, B.; Dietz, R.J.B.; Kohlhaas, R.B.; Göbel, T.; Schell, M.; Alcer, D.; Semtsiv, M.; Masselink, W.T. Iron doped InGaAs: Competitive THz emitters and detectors fabricated from the same photoconductor. *J. Appl. Phys.* **2017**, *121*, 053102. [CrossRef]
43. Kocic, N.; Wichmann, M.; Hochrein, T.; Heidemeyer, P.; Kretschmer, K.; Radovanovic, I.; Mondol, A.S.; Koch, M.; Bastian, M. Lenses for terahertz applications: Development of new materials and production processes. *AIP Conf. Proc.* **2014**, *1593*, 416–419. [CrossRef]
44. Harris, Z.B.; Virk, A.; Khani, M.E.; Arbab, M.H. Terahertz time-domain spectral imaging using telecentric beam steering and an f-θ scanning lens: Distortion compensation and determination of resolution limits. *Opt. Express* **2020**, *28*, 26612–26622. [CrossRef]
45. Kong, D.Y.; Wu, X.J.; Wang, B.; Gao, Y.; Dai, J.; Wang, L.; Ruan, C.J.; Miao, J.G. High resolution continuous wave terahertz spectroscopy on solid-state samples with coherent detection. *Opt. Express* **2018**, *26*, 17964–17976. [CrossRef] [PubMed]
46. Wachulak, P.W.; Torrisi, A.; Bartnik, A.; Adjei, D.; Kostecki, J.; Wegrzynski, L.; Jarocki, R.; Szczurek, M.; Fiedorowicz, H. Desktop water window microscope using a double-stream gas puff target source. *Appl. Phys. B* **2015**, *118*, 573–578. [CrossRef]
47. Wachulak, P.W.; Torrisi, A.; Bartnik, A.; Węgrzyński, Ł.; Fok, T.; Fiedorowicz, H. A desktop extreme ultraviolet microscope based on a compact laser-plasma light source. *Appl. Phys. B* **2016**, *123*, 25. [CrossRef]
48. Di Fabrizio, M.; D'Arco, A.; Mou, S.; Palumbo, L.; Petrarca, M.; Lupi, S. Performance Evaluation of a THz Pulsed Imaging System: Point Spread Function, Broadband THz Beam Visualization and Image Reconstruction. *Appl. Sci.* **2021**, *11*, 562. [CrossRef]
49. Park, S.J.; Yoon, S.A.N.; Ahn, Y.H. Dielectric constant measurements of thin films and liquids using terahertz metamaterials. *RSC Adv.* **2016**, *6*, 69381–69386. [CrossRef]

**Disclaimer/Publisher's Note:** The statements, opinions and data contained in all publications are solely those of the individual author(s) and contributor(s) and not of MDPI and/or the editor(s). MDPI and/or the editor(s) disclaim responsibility for any injury to people or property resulting from any ideas, methods, instructions or products referred to in the content.

MDPI AG
Grosspeteranlage 5
4052 Basel
Switzerland
Tel.: +41 61 683 77 34

*Sensors* Editorial Office
E-mail: sensors@mdpi.com
www.mdpi.com/journal/sensors

Disclaimer/Publisher's Note: The title and front matter of this reprint are at the discretion of the Guest Editors. The publisher is not responsible for their content or any associated concerns. The statements, opinions and data contained in all individual articles are solely those of the individual Editors and contributors and not of MDPI. MDPI disclaims responsibility for any injury to people or property resulting from any ideas, methods, instructions or products referred to in the content.

www.ingramcontent.com/pod-product-compliance
Lightning Source LLC
LaVergne TN
LVHW070002100526
838202LV00019B/2613